BIOLOGICAL PEST CONTROL

THE GLASSHOUSE EXPERIENCE

Edited by

N. W. HUSSEY & N. SCOPES

CORNELL UNIVERSITY PRESS
ITHACA, NEW YORK

First published in 1985 by Cornell University Press

Library of Congress Cataloging in Publication Data

Main entry under title:
Biological Pest Control
Bibliography: p.
Includes index.
1. Greenhouse plants - Diseases and pests.
2. Insect pests - Biological control.
3. Plant mites - Biological control.

I. Hussey, N. W. (Norman Waltham)
II. Scopes, Nigel E. A. (Nigel Eric Anthony)
SB936.B56 1985 635' 04996 85-7896

ISBN 0-8014-1835-6

Originally published in the U.K. by Blandford Press Ltd.

Typeset by August Filmsetting, Haydock, St. Helens.
Printed and bound in Great Britain.

CONTENTS

LIST OF CONTRIBUTORS

Bassett, P. (Agricultural Development and Advisory Service, Starcross, Exeter, Devon, UK)
Campbell, C.A.M. (East Malling Research Station, Maidstone, Kent, UK)
Copland, M.J.W. (Department of Biological Sciences, Wye College, University of London, Ashford, Kent, UK)
Easterbrook, M. (East Malling Research Station, Maidstone, Kent, UK)
Fournier, D. (Institut National de la Recherche Agronomique, Station de Zoologie et de Lutte Biologique, 37 Boulevard du Cap, F-06602, Antibes, France)
Gould, H.J. (Ministry of Agriculture Fisheries & Food, Harpenden Laboratory, Hertfordshire, UK)
Hall, R.A. (Glasshouse Crops Research Institute, Littlehampton, West Sussex, UK)
Hassan, S.A. (Institut fur biologische Schadlingsbekampfen, D-6100 Darmstadt, West Germany)
Helyer, N. (Glasshouse Crops Research Institute, Littlehampton, West Sussex, UK)
Hussey, N.W. (formerly Glasshouse Crops Research Institute, Littlehampton, West Sussex, UK)
Ibrahim, A.G. (Department of Plant Protection, Universiti Pertanian, Malaysia)
Jarrett, P. (Glasshouse Crops Research Institute, Littlehampton, West Sussex, UK)
Lipa, J.J. (Institute of Plant Protection, Poznan, Poland)
Ledieu, M. (Glasshouse Crops Research Institute, Littlehampton, West Sussex, UK)
Lisansky, S.G. (Microbial Resources Ltd, 10 Deacon Way, Reading, Berkshire, UK)
Lyth, M. (East Malling Research Station, Maidstone, Kent, UK)
Markkula, M. (Institute of Pest Investigation, Agricultural Research Centre, Jokioinen, Finland)
Millot, P. (Institut National de la Recherche Agronomique, Station de Zoologie et de Lutte Biologique, 37 Boulevard du Cap, F-06602, Antibes, France)
Oomen, P.A. (Institut for biologische Schadlingsbekampfen, D-6100 Darmstadt, West Germany)
Panis, A. (Institut National de la Recherche Agronomique, Station de Zoologie et de Lutte Biologique, 37 Boulevard du Cap, F-06602, Antibes, France)
Pickford, R. (Humber Growers Marketing Organisation, Brough, Yorkshire, UK)

Pralavorio, M. (Institut National de la Recherche Agronomique, Station de Zoologie et de Lutte Biologique, 37 Boulevard du Cap, F-06602, Antibes, France)
Rabasse, J.M. (Institut National de la Recherche Agronomique, Station de Zoologie et de Lutte Biologique, 37 Boulevard du Cap, F-06602, Antibes, France)
Rombach, M.C. (CentraalBureau voor Schimmelcultures, Baarn, the Netherlands)
Samson, R.A. (CentraalBureau voor Schimmelcultures, Baarn, the Netherlands)
Saynor, M. (Agricultural Development and Advisory Service, Starcross, Exeter, Devon, UK)
Scopes, N.E.A. (formerly Glasshouse Crops Research Institute, Littlehampton, West Sussex, UK)
Soeprapto, W. (Department of Biological Sciences, Wye College, University of London, Ashford, Kent, UK)
Stenseth, C. (Plant Protection Institute, As-NLH, Norway)
Tiittanen, K. (Institute of Pest Investigation, Agricultural Research Centre, Jokioinen, Finland)
Tingle, C.D. (Department of Biological Sciences, Wye College, University of London, Ashford, Kent, UK)
Wardlow, L.R. (Agricultural Development and Advisory Service, Wye Sub-Centre, Ashford, Kent, UK)
Woets, J. (Glasshouse Crops Research and Experiment Station, Naaldwijk, the Netherlands)
Wyatt, I.J. (Glasshouse Crops Research Institute, Littlehampton, West Sussex, UK)

FOREWORD

The international scientific co-operation fostered by the Working Parties of the West Palaearctic Regional Section of the International Organization for Biological Control has made an immense contribution to the development of rational pesticide usage in Western Europe. While the results of these endeavours normally appear in scientific journals they are usually concerned with specific topics so that an overview can be achieved only by a wide perusal of the literature. This is certainly the case with the activities of the group concerned with Integrated Control in Glasshouses. The Council of the Section therefore requested that an attempt should be made to collate the information in a form which would provide users with practical information and the uninformed with an account of one of the outstanding examples of successful integrated control.

We are indebted to the authors for their readiness to provide texts and photographs, to the staff of the Glasshouse Crops Research Institute for their many contributions and especially to Mrs Sue Bewsey who painstakingly undertook the typing of the texts.

N.W. Hussey
N.E.A. Scopes

Commonwealth Institute of
Biological Control
Silwood Park
Ascot
Berkshire, UK

ACKNOWLEDGEMENTS

COLOUR PLATES: P. Bassett/ADAS: 23; British Museum (Natural History): 6; M. Copland/Wye College: 17, 18, 19, 21; Crown Copyright: 15; Glasshouse Crops Research Institute: 1, 2, 5, 8, 9, 11, 12, 13, 14, 16, 20, 22, 25, 26, 27; R. Lindquist, Wooster, Ohio: 10; D.G. Mason/Coutts & Co.: 28; R.A. Samson/CBS, Baarn: 3, 4.

BLACK AND WHITE PHOTOGRAPHS: P. Basset/ADAS: pp. 93, 94; Coutts & Co.: p. 186; Glasshouse Crops Research Institute: pp. 8, 9 (bis), 35 (bis); 37, 43, 44, 46, 47, 49, 54, 59 (bis); 63, 67, 100, 206, 207, 208, 209; R.A. Samson/CBS, Baarn: p. 39 (bis); M. Saynor/MAFF: p. 85; N. Scopes/GCRI: p. 201; Tate & Lyle Ltd/GCRI: pp. 213, 214; I. Wyatt/GCRI: p. 71.

INTRODUCTION

The greenhouse provides an almost ideal environment for crop production, providing warmth, food and water to plants while protecting them from the extremes of weather (frosts, wind and draught etc.). Such conditions also favour insect development and reproduction.

The multiplication rates of some pests are prodigious, for instance Wyatt & Brown (1977) showed that *Aphis gossypii* could multiply by as much as 22.9 × each week in the laboratory; however, in practice, an increase of 10 × is more usual in greenhouses. Alternatively, one has only to see the hundreds of eggs produced by a single scale insect (Plate 20) to realize the enormity of the problems posed by some pests.

This rapid multiplication of greenhouse pests necessitates routine spray programmes. These intensive programmes have led to the

FIGURE 1 'Reproduction'

selection of 'super bugs', i.e. strains that are resistant (tolerant) to the active ingredients which give toxicity to pesticides. It was not uncommon for more than 20 sprays to be applied to a single cucumber crop and such intense programmes exerted a powerful pressure in favour of individuals possessing genes for resistance. This process is a most interesting example of Darwinian selection. The mechanisms conferring resistance usually take the form of enzyme systems which detoxify the active toxic moiety. It is not surprising that the rapidly reproducing red spider mite was the first greenhouse pest to develop resistance, strains tolerant to azobenzene appearing in 1949. More recently, leaf-miners, aphids and whiteflies have all shown that they also possess genes conferring resistance to the wide range of chemicals applied to control them. Inconveniently, such new 'strains' appear more rapidly than Man's ingenuity can develop new compounds. The saying that 'Resistance is a one-way street – there is no going back' is very true.

FIGURE 2 'Resistance'

The combination of rapid multiplication rates and resistance to pesticides creates a formidable obstacle to the continued use of chemical pesticides. Such problems, however, are made even more complex if one accepts the opinion, expressed by some growers, that each pesticide spray reduces yield. One reputable cucumber-grower has reported a 25% yield increase by using biological control to control red spider mites rather than his normal routine of 23 sprays. The complexities of such yield losses, apparently due to direct toxicity to plants, are not, as yet, understood.

The problems of controlling greenhouse pests are further aggravated by the cultural changes that have taken place in response to economic pressures. Tomatoes, for instance, are widely grown in isolated sub-

FIGURE 3 'Shelter from pesticides'

strates to avoid the necessity to sterilize the border soil. This cessation of annual sterilization allows both pests and diseases to overwinter and attack following crops while the nearly continuous monoculture may perpetuate resistant strains within the greenhouse.

In the face of increasing labour costs, advances in crop protection technology by using labour-saving, low-volume application systems have been made to achieve financial savings. Thermal fogging, for instance, provides a quick and convenient method of applying chemicals but, unfortunately, only to the upper leaf surfaces. Unless the limitations of such systems are fully appreciated, it seems reasonable to assume that resistance problems will be aggravated by the incomplete cover so that additional applications may be needed.

In addition to the insect and plant problems already mentioned, it is necessary to make many spray treatments in the evenings so as to reduce phytotoxic damage by sunlight. Furthermore, many chemicals are poisonous, making it essential for operators to wear cumbersome, protective clothing for their safety.

It is hardly surprising that, in the face of these limitations, biological methods of pest control have considerable appeal since they are regarded as safe and natural. However, living controls have to be used intelligently with a sound knowledge of their biology. Several natural enemies are now used to control the important pests on a wide range of crops throughout Europe. They operate within integrated programmes which harmonize the use of natural enemies with pesticides to control minor pests and diseases.

This book describes the biology of both the important greenhouse pests and their natural enemies. It outlines the history and development of these systems, while providing sufficient practical detail to enable readers to implement the techniques for themselves.

REFERENCES

Wyatt, I.J. & Brown, S.J. (1977) The influence of light intensity, daylength and temperature on increase rates of four greenhouse aphids. *J. Appl. Ecol.* **14**: 379–399.

1 HISTORY OF BIOLOGICAL CONTROL IN PROTECTED CULTURE

1.1 WESTERN EUROPE

N.W. Hussey

Perhaps the earliest reference to natural pest control in closed environments was outlined by J.C. Loudon (1850) in his *Encyclopaedia of Gardening* (Book III, Pt III, p. 819) when he suggested that 'a toad kept in a mushroom house will eat worms, ants and other insects, but to most people the idea would be disgusting of a toad crawling over anything intended for the table'.

The next pioneer was G. Fox-Wilson, entomologist at the Royal Horticultural Society Gardens at Wisley in Surrey who found the whitefly (*Trialeurodes vaporariorum*) being attacked by a native parasite, *Encarsia partenopea.* Although maintained for a few years and distributed to certain members, use of the parasite found no permanent place in glasshouse practice, probably because it was ineffective under high summer temperatures.

It was not until 1926, when a tomato-grower drew the attention of Mr E.R. Speyer to black scales among the normally white host scales, that the exotic parasite, *Encarsia formosa*, was formally identified (Speyer, 1927) and a new era of biological control began. As well as studying the biology of this parasite, Speyer developed systems of mass production which led, by 1935, to the annual production of 1½ million wasps and their distribution to seven countries around the world. At that time, the introduction techniques were based on the use of a fixed number of parasites per unit area (9/m²), as determined by McCleod (1938) in Canada. This system, although simple, must have given variable results though there were many claims that 'control' was satisfactory. However, it should be remembered that the economics of crop protection were less demanding in the 1930s. For instance, tomato-growers would consider that a 165 tonne/ha crop would provide an adequate return whereas, today, twice that yield would be needed to give a realistic margin above dramatically increased production costs. Production of *Encarsia* continued at the Cheshunt Experiment Station, although at a reduced level during the War years, until 1949. By that time, a new

NATIONAL FARMERS' UNION

WORTHING AND WEST SUSSEX GROWERS.

Telephone: Worthing 858.

Ivy Arch Road,
Worthing.

16th October, 1933.

A

LANTERN LECTURE

on

THE USE OF PARASITES IN THE CONTROL OF GLASSHOUSE PESTS

will be given by

MR. EDWARD R. SPEYER, M.A.,

at the

CENTRAL HOTEL, WORTHING,

on

THURSDAY, OCTOBER 26th, 1933,

at 7.30 p.m.

Mr. Speyer, the entomologist of the Research Station, Cheshunt, is well known to many growers in this district. He has prepared a very interesting and instructive lecture which will be illustrated with an entirely new series of lantern slides, and which will be delivered in non-technical language. The following is a synopsis of the lecture :-

1. Increasing difficulties in the control of major pests of tomatoes and cucumbers by methods of spraying and fumigation.

2. The nature of insect parasites of these pests.

3. The White-Fly Parasite; its life history, method of distribution and the economic benefit derived therefrom.

4. Parasites of the Tomato-moth-caterpillar :
   (a) the Chalcid,
   (b) the Ichneumon,
   Their life histories and economic status.

5. A predatory enemy of the Red-Spider Mite - the Gall Midge. The possibility of its use in the control of the Mite on tomatoes and peaches.

The Education and Entertainments Committee extend a cordial invitation to all members and their foremen or other interested persons.

FIGURE 4 Notice of lecture by Mr. E. Speyer (Cheshunt Experimental Station) to Worthing growers.

generation of synthetic insecticides, headed by DDT, had become available to growers and it was decided that production should cease. Hence, when the staff of Cheshunt moved to the new Glasshouse Crops Research Institute (GCRI) in 1956, no natural enemies remained in culture.

While the progress in biological control of whiteflies was unfolding, Speyer sought to widen its impact to other glasshouse pest problems and it is, therefore, interesting to record that I have the original notes of a lecture which he presented to Worthing growers in 1935. In this talk, he raised the possibilities of using the ichneumonid parasites, *Pimpla instigator*, to control tomato moth (*Lacanobia oleracea*) and a midge (*Therodiplosis persicae*) to reduce populations of spider mite. No interest was aroused in natural control of tomato moth, largely because the widespread introduction of steam sterilization of soil prevented overwinter-survival of the pupae. However, since the mid-1970s, as hydroponic systems and the use of peat-bolsters have become popular, the pest has increased in numbers so that serious outbreaks, particularly in the Channel Islands and the Clyde Valley in Scotland, are now common. Similarly, while no attempt was made to develop the midge predator of spider mites, widespread use of *Phytoseiulus*, and hence reduction in chemical usage, has allowed the midge to survive and demonstrate its potential. Indeed, Foster has observed that, in the Clyde Valley, *Therodiplosis*, which finds its way naturally into greenhouses, is responsible for a greater proportion of the reduction in mite numbers than *Phytoseiulus*.

However, biological control attracted little further attention until Bravenboer (1960) published his important paper on chemical and biological control of *Tetranychus urticae* Koch. This paper, based on studies for his doctorate thesis, later led Bravenboer to become a pioneer of the Organisation Internationale de Lutte Biologique (OILB) and for 12 years was convener of the Working Party on Integrated Control in Glasshouses. He drew attention to the fact that, after the introduction of DDT in 1945, serious outbreaks of spider mites occurred and that, although the use of organophosphorous (OP) compounds solved the problem temporarily, resistance steadily increased the number of applications necessary until total failure occurred in numerous cases during the early 1950s.

Bravenboer found that, by 1956, the coccinellid, *Stethorus punctillum*, occurred at the end of each season on peaches and nectarines on 60% of the nurseries examined, despite the routine use of OP acaricides. This monophagous predator, though reproducing more slowly than its mite host, was shown to be an important predator as each beetle consumes some 100 mites daily.

Surveys of glasshouse fruit-growing districts in the Netherlands in

FIGURE 5 Members of OILB Working Group on Integrated Control in Glasshouses in Darmstadt, July 1982.

1956 showed that the mite, *Typhlodromus longipilus* Nesbitt, was common on 75% of the holdings. This predatory mite develops twice as rapidly as the spider mite, although its fecundity is half that of *Stethorus* and it eats only about 10 prey adults daily.

However, both predators are density-dependent enemies capable of regulating red spider mite populations and so only one selective acaricide application was needed to reduce the prey density sufficiently for both predators, acting in concert, to provide adequate control without resistance problems. The full significance of Bravenboer's work was never appreciated for, in 1960, Dr Dosse had found the now legendary *Phytoseiulus persimilis* (=*reigeli*) on a consignment of orchids received in Germany from Chile. He distributed this predator to interested workers in several European countries and all agreed that it was a most effective predator (Bravenboer & Dosse, 1962), although many felt that it was not suitable for commercial use as excessive monitoring and manipulation was required. However, at the GCRI in the UK, an important series of studies established the economic damage thresholds for red spider mite and whitefly which concentrated attention on the need to maintain a certain photosynthetic area on the plants. By inference, this demonstrated the need for pest populations to oscillate uniformly throughout the greenhouse. Based on this philosophy, the 'pest-in-first' approach was pioneered which, while rarely used commercially, did create confidence among growers and the methods now widely adopted are based on this somewhat heretical philosophy.

The ideas, until then developed in small experiments, were expanded in 1966 to trials in some 40 large, commercial cucumber-houses at Rochford's Nurseries at Slough, near London Airport. All the necessary predators were collected individually so that every alternate plant received a small gelatine capsule containing 2 predators and 5 red spider mites. This approach was spectacularly successful and a very serious spider mite problem was solved. Indeed, the pest was almost eliminated from treated houses, so that the regular invasion of newly planted crops by ex-diapause mites did not occur subsequently. More significantly, the 'release' of the crop from the prophylactic programme of more than 20 acaricide applications per season led to spectacular increases in plant growth, accompanied by crop increases of 25%.

Despite this success, it was realized that further progress depended on the need to harmonize the control of all the other pests and diseases on cucumbers with this predatory mite (Hussey, 1968).

Although, as mentioned earlier, practical use of the whitefly parasite, *Encarsia formosa*, had ceased in 1949, the species was known to have persisted in a number of hot-houses within Botanic Gardens so that, in 1967, the GCRI was again able to set up cultures from a few parasitized scales obtained from Cambridge.

During the development phase of biological control, the group at GCRI assumed that they would ultimately be responsible for the mass production of natural enemies but, in the event, the authorities in London would not agree to such a dilution of research effort and so attempts were made to attract the interest of individuals in the horticultural industry.

The first of these, Mr E. Hubbard, a cucumber-grower, of Springfield Nursery, Waltham Cross, was the first to produce *Phytoseiulus* for sale and he showed the advantages of rearing spider mites in a hot, dry environment (an artificially lit boiler house). Enormous numbers could be rapidly reared in a very confined space at minimal cost. At about the same time, Mr Mark Savage, a teacher and pesticide company representative, set up in business for a short period but the individual with the most lasting effect on the British scene was Mr Ivan Worrall, who had been the foreman in charge of pest control procedures at the Rochford Nursery in Slough. Once the advantages of biological control had been confirmed by the field trials mounted jointly by the Agricultural Development and Advisory Service (ADAS) and GCRI, a small production unit was set up there to provide for the 8.0 ha of cucumbers grown by that company. Later, when the nursery was moved, Mr Worrall purchased the rearing unit and set up in business as Natural Pest Control, a firm which remains in business today, although it moved to a new production site in Sussex in 1979. Throughout this period, Mr Worrall's enterprise was sustained by the confidence which growers placed in him as, unlike the earlier producers, he provided a technical advisory service which involved visits to nurseries to ensure that the respective natural enemies had been successfully established. During the 1970s, several other enthusiasts – Mr Woodcock of Plymouth; Kent Country Nurseries, a firm with which A. Ludlam (an advisory officer who had helped to develop some of the management techniques and who was later killed in a private flying accident) was associated; and S. Legowski, an advisory entomologist, who later joined the Sandoz Chemical Co. in Switzerland – all produced *Phytoseiulus* and *Encarsia* for sale.

No doubt the most significant event occurred in 1969, when Mr Koppert, a cucumber-grower and father of Peter and Paul, who now run the largest commercial production unit, visited the GCRI and was persuaded to enter the natural enemy business. His letter (Figure 6) represents a milestone in the development of biological control in Western Europe. This company made major improvements in the production techniques by pioneering the sale of mixed cultures of *Phytoseiulus* and spider mites in bran and also perfected methods of removing whitefly scales parasitized by *Encarsia* from the host plant and selling them 'glued' to cards provided with an artificial food-source and a convenient means for hanging in the crop. This advance had important implications for international sales.

While this development which ultimately led to that firm capturing 70% of the market share in Western Europe, was occurring in the Netherlands, the speed of acceptance of the technique in the UK was obviously hampered by lack of confidence on the part of growers. One firm, Perifleur of Rustington, Sussex, attempted to overcome this handicap by recruiting qualified staff who sold and provided supervised control by contract. This far-sighted scheme failed because it proved uneconomic to maintain such expensive staff in the field with only a seasonal cash-flow – there being little demand for advice from August to January.

Perhaps not surprisingly, the logical counter to lack of faith in commercial suppliers was the setting up, by certain growers, of rearing units initially designed to provide sufficient material for a few large nurseries only. Foremost amongst these was the Humber Growers Marketing Organization near Hull, for whom Robert Pickford became a highly effective technician. At the same time, another firm, Bunting & Sons of Colchester, used a similar approach which was later expanded to serve other growers and, by 1983, an entry into the potentially attractive market in Spain was provided by the large concentration of plastic-house production in that country.

Commercial biocontrol in Norway began with the use of *Phytoseiulus* in 1971. For the first 3 years, the predator was imported from Finland but production began at LOG, Oslo, in 1974. An OP-resistant strain has

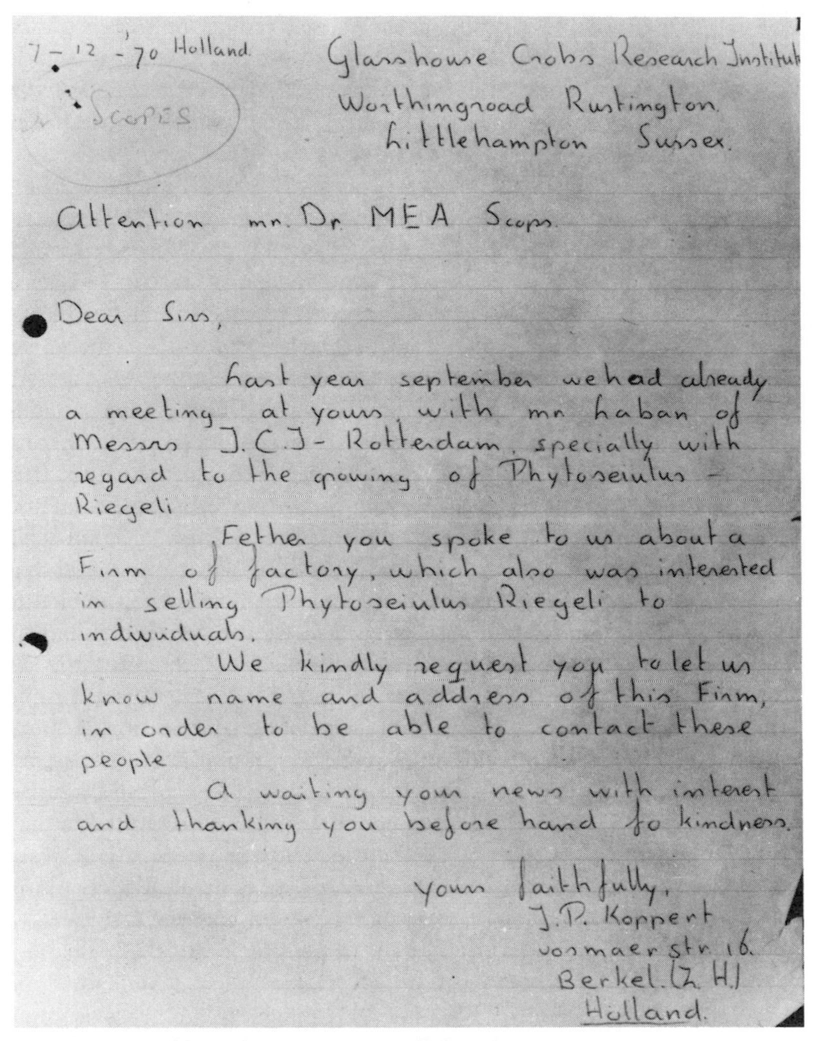
7-12-70 Holland.

Scopes

Glasshouse Crops Research Institut
Worthingroad Rustington.
Littlehampton Sussex.

Attention mr. Dr. MEA Scops.

Dear Sirs,

Last year september we had already a meeting at yours, with mr Laban of Messrs J.C.J- Rotterdam, specially with regard to the growing of Phytoseiulus - Riegeli

Fether you spoke to us about a Firm of factory, which also was interested in selling Phytoseiulus Riegeli to individuals.

We kindly request you to let us know name and address of this Firm, in order to be able to contact these people

A waiting your news with interest and thanking you before hand fo kindness.

Yours faithfully,
J.P. Koppert
vosmaer str 16.
Berkel. (Z.H.)
Holland.

FIGURE 6 Original letter from Mr J. Koppert (father of Peter Koppert) to GCRI.

been used since 1978. One large cucumber-grower produces his own *P. persimilis*.

The growers have always been very interested in the biocontrol method but the results have not always been so good as desired. Both research, to adjust the use of *P. persimilis* to Norwegian conditions, and training of the growers has been important for successful use of the method.

In Norway, both cucumbers and tomatoes are planted in January–March when outside temperatures are very low. This creates low relative humidities in the greenhouses which reduce predator efficiency until sufficient foliage is present to increase atmospheric moisture. A combination of large numbers of post-diapausing females and low humidity can upset biocontrol. A change from patch treatment to uniform introduction has improved the situation.

Encarsia formosa was used in some places before World War 2 but, while commercially organized distribution started in 1972, it was not used widely until 1975. The parasite is sent by post directly from a producer in England to growers in Norway, the transit taking 2–5 days. This operates effectively as the producer has evolved a system of production which ensures that any unparasitized scales emerge as whitefly 8–9 days before the parasite. This allows a 7-day transit period.

Having sketched some of the commercial developments which became successful through the close association between both research and advisory staff and practical growers, it is timely to resume a consideration of the technical advances. While the outlines of an interaction between pest and natural enemy could be determined in small-scale research experiments, real progress depended on commercial experience. Hence, the experimentation proceeded step by step in closely monitored trials on commercial nurseries. Hence, though scientists might wince at the lack of classical controlled experimentation, the confidence which such large-scale trials engendered encouraged rapid uptake of the methods. The pressures brought by the horticultural industry led to the development of several different methods, i.e. 'pest-in-first', 'dribble' and 'bankers', which were designed to achieve the same technical objective – a host:natural enemy ratio so much in the latter's favour that control must ensue – and yet appeal to growers with different degrees of confidence in the concept of biological control. A particularly important role in this development work in the UK was played by a team of advisory entomologists led by H.J. Gould (including N. French, J. Legowski, P. Simmonds, L. Wardlow, H. Woodville and J. Williams) and D. Green, a horticulturalist in the Lee Valley, who had a great personal influence on growers in that area. This team effort provided a unique example of co-operation in research and development which was largely responsible for the rapid exploitation of the new methods in the UK during the 1970s.

Having devised commercially effective biological control techniques for both whitefly and red spider mite, attention turned to the problems of harmonizing the chemical methods used for plant diseases and minor pests with parasites and predators. M. Ledieu and N. Helyer at GCRI have played a major role in this work which proceeded within the framework of another OILB Working Party convened by S. Hassan of Darmstadt and to which Drs Coulon (France), Jacobsen and Kirknel have also made significant contributions.

The results of all this endeavour were synthesized in two practical guides published by and obtainable from the GCRI, Littlehampton, Sussex, UK, which elaborate fully integrated programmes for both cucumbers and tomatoes.

In the Netherlands, it had been concluded that problems with *Thrips tabaci* on cucumbers precluded the use of whitefly parasites whereas, in the UK, excellent results were claimed. Subsequent studies by the Commonwealth Institute of Biological Control (Delemont in Switzerland) revealed that, where cucumbers were grown on traditional manure beds or straw bales irrigated with overhead spray lines, the excess surface water encouraged natural control of pupating thrips by the fungus, *Entomophthora thripidum* (Carl, 1975). However, devastating outbreaks of thrips caused serious problems when rock-wool culture increased and, following a special OILB meeting at GCRI in 1978, different approaches were pursued in different countries. In the Netherlands, Ramakers (Naaldwijk) studied the predatory mites, *Typhlodromus cucumeris* and *T. mackenziei*, perfecting a mass production system which was exploited by Koppert BV. Field experience has shown that these predators provide a rather slow control. The accidental observation by Hassan (Darmstadt) that the fungicide, pyrazophos, was lethal to *T. tabaci* has stimulated much interest. However, the most rapid and selective method was devised and commercialized as a result of co-operative work between Humber Growers and the GCRI. This method stemmed from attempts to replace soil treatment with γ-HCH by other soil-applied pesticides. Since, in rock-wool cultures, the floor is covered with plastic sheeting, polybutenes were tested as a physical trap. Though effective, the sticky surface was regarded as impractical but later a mixture of polybutenes and deltamethrin produced excellent results lasting 10–12 weeks following a single application and this material is now marketed as Thripstick®.

During the 1970s, many workers encountered difficulties in controlling aphids, especially where resistance had developed to the carbamates, aldicarb and pirimicarb. Much research was done on coccinellids in France, Finland and the UK and on chrysopids in Norway, the USA and the UK. Whilst it was clearly demonstrated that control could be achieved, the fact that no successive generations developed because female predators would oviposit freely only in the presence of an unacceptably high pest population made these aphid predators uneconomic. However, more success stemmed from work on *Aphidius matricariae* by Wyatt & Scopes at GCRI and by Lyon on *Diaeretiella rapae* at Antibes, for both gave excellent control of the ubiquitous *Myzus persicae*. Indeed, GCRI's *Grower's Bulletin* No. 2 on 'Rearing your own parasites and predators' included appropriate instructions. However, the presence of a complex of aphid pests on different crops (*M. persicae* on tomatoes and chrysanthemums, *Macrosiphum euphorbiae* on peppers and cucumbers, *Aphis gossypii* on cucumbers and chrysanthemums, *Macrosiphum rosae* on roses, *Aulocorthum solani* and *Brachycaudus helichrysi* on chrysanthemums) called for a non-specific natural enemy.

Markkula *et al.* (1979) started to work on the midge, *Aphidoletes aphidimyza*, in 1971, at the same time as other workers in Germany and Leningrad. All research confirmed the remarkable efficacy of this midge but the Finnish workers were the first to conduct successful field experiments using a commercially practicable method of production and application. The midge is introduced as pupae held in damp sand and, in the presence of aphids, develops several successive generations before decreasing autumnal day length induces winter diapause. Furthermore, where no sterilization methods are adopted, midges reappear the following spring without further introductions.

Despite developments in the field of pathogens, *Aphidoletes* may yet be developed to play a major role in biological pest control, especially where, as on chrysanthemums, a complex of aphid pest species is often present simultaneously.

When Hussey joined the GCRI in 1956, he isolated a fungus, *Verticillium lecanii*, attacking a dense population of whiteflies on glasshouse cucumbers in North Devon. Having made a preliminary and, as it turned out, incorrect, analysis of its potential (Hussey, 1958), he maintained it in culture for more than 10 years when he was joined by Dr Hall, who clearly showed its efficiency as a control for all the leaf-feeding aphids on chrysanthemums as well as for whiteflies on cucumbers. Subsequently, these strains were developed and marketed by Tate & Lyle as Vertalec® and Mycotal® respectively. Yet another strain has been shown to be pathogenic to thrips.

Nearly all attempts at biological control have been hindered by the need to control caterpillar pests with broad-spectrum insecticides. This difficulty has recently been compounded by the development of the synthetic pyrethroids, which render foliage poisonous to parasites for as long as 12 weeks after application. Attention has therefore focussed on commercial preparations of *Bacillus thuringiensis.* Dr Burges and his colleagues at GCRI paid particular attention to the application of this material through fogging machines and concluded that, while this was feasible with pulse-jet foggers (Pulsfog®, Swingfog®, Dynafog®), it should not be attempted with exhaust type machines, due to the high temperatures developed in the barrel. Better results were usually obtained by HV applications in the order of 400 l/ha. It is important to vary the rate according to the species involved as follows: for *Mamestra brassicae* use 0.6% *Bacillus thuringiensis* (8 kg in 40 l VK2/ha) while other species can be controlled with half this quantity.

Although no safety clearance has yet been sought, successful control of *Laconobia oleracea* by granulosis virus has been claimed by Foster & Crook (1983) and of *Spodoptera exigua* by Dutch workers (Vlak *et al.*, 1982).

These endeavours by the membership of the OILB Working Party on 'Integrated Control in Glasshouses' has led to the regular use of the methods on more than 1000 ha of commercial glass in the 10 years since 1968. More than 95% of the operations have been classified as successful by the commercial growers and, as the techniques become adapted to the rather different cultural methods used in warmer Mediterranean countries, the rearing companies can look forward to a potential £10 million annual turnover. It has been a unique international effort which is now extending to other intensively grown crops that have promised exceptional results in cases where growers are willing to invest some of their cost savings in reduced spraying and in monitoring the pest and disease situation on their crops in order to ensure timely biological interventions.

An interesting new development in the use of biological control in the UK has occurred on plants used for internal decor in public buildings. Frequently, these plants are infested with mites, scales and mealybugs in situations where the use of pesticides is undesirable or even impossible. Where natural enemies and pathogens have been exploited it has created considerable public interest greatly to the long-term benefit of natural control techniques.

REFERENCES

Bravenboer, L. (1960) De Chemische en Biologische bestrijding van de Spintmijt *Tetranychus urticae* Koch. *Publikatie Proefstn Groenten-en Fruiteelt onder Glas te Naaldwijk* No. 75: 85 pp.

Bravenboer, L. & Dosse, G. (1962) *Phytoseiulus riegeli* Dosse als Predator einiger Schadmilben aus der *Tetranychus urticae* gruppe. *Entomologia Exp. Appl.* **5**: 291–304.

Carl, K.P. (1975) An *Entomophthora* sp. pathogenic to *Thrip* sp. and its potential as a biological agent in glasshouses. *Entomophaga* **20**: 381–388.

Foster, G.N. & Crook, N.E. (1983) A granulosis disease of the tomato moth *Lacanobia oleracea. Bull. OIBC/WPRS Working Group on Integrated Control in Glasshouses, Darmstadt* **6** (3): 163–166.

Hussey, N.W. (1958) Notes on a fungus parasite on greenhouse whitefly. *Pl. Pathol.* **7**: 71–72.

Hussey, N.W. (1968) Prospects for integrated control in protected cultivation. *Chem. Indust.* April 20: 498–502.

Hussey, N.W., Parr, W.J. & Gould, H.J. (1965) Observations on the control of *Tetranychus urticae* Koch on cucumbers by the predatory mite *Phytoseiulus riegeli* Dosse. *Entomologia Exp. Appl.* **8**: 271–281.

McLeod, J.H. (1938) The control of greenhouse whitefly in Canada by *Encarsia formosa. Scient. Agric.* **18**: 529–535.

Markkula, M., Tiittanen, K., Hamalainen, M. & Forsberg, A. (1979) The aphid midge *Aphidoletes aphidimyza* and its use in the biological control of aphids. *Annls Agric. Fenn.* **45**: 89–98.

Parr, W.J. & Hussey, N.W. (1967) Biological control of red spider mites on cucumbers: effects of different predator densities at introduction. *Rep. Glasshouse Crops Res. Inst.* (**1966**): 135–139.

Ramakers, P. (1978) Possibilities for biological control of *Thrips tabaci* in glasshouses. *Meded. Rijksfac. Landb. Gent* **43**: 463–469.

Speyer, E.R. (1927) An important parasite of the greenhouse whitefly. *Bull. Ent. Res.* **17**: 301–308.

Vlak, J.M., den Belder, E., Peters, D. & van de Vrie, M. (1982) Bekämpfung eines eingeschleppten schädlings *Spodoptera exigua* in Gewachshausen mit den autochtonen virus. *Meded. Proefst Aalsmeer* No. 81: 1005–1016.

1.2 EASTERN EUROPE

Jerzy J. Lipa

Studies on the biological control of pests and diseases in Eastern Europe have been both broad and intensive. However, the scale of practical application to protect greenhouse plants varies between countries. The main constraint is that in no country is there a commercial supply of biotic agents. Therefore, each interested state farm or private owner has to rear his own biotic agents. In spite of this constraint, the area on which these agents are used has steadily increased. Especially impressive figures refer to the USSR where, in 1981, 6150 ha of greenhouse crops were protected with biological methods, of which 2666 ha were protected against *Tetranychus* with *Phytoseiulus persimilis.*

1.2a **SPIDER MITE CONTROL** Spider mites (*Tetranychus urticae* and *T. cinnabarinus*) are the most common and important pests on several greenhouse crops. Biological control by a predatory mite, *Phytoseiulus persimilis*, is used in many countries. Some studies have also been conducted on the use of fungi to control spider mites but they are still in a preliminary stage.

Phytoseiulus persimilis This predator mite was introduced from Canada to the USSR in 1963 (Begljarov *et al.*, 1964) from England to Poland in 1965 (Lipa *et al.*, 1967) and, some years later, to Roumania, Bulgaria, Czechoslovakia and the German Democratic Republic. At present, this predator is used on a relatively large scale in the USSR and on a much smaller area in Poland, Roumania and Bulgaria.

Increasing costs of pesticides and their application, as well as mite resistance and pesticide residues, have stimulated interest in the use of *P. persimilis.* In the USSR, in 1967, the predator was used on an area of 1.6 ha, in 1968 on 22.8 ha; in 1971 on 117.8 ha; in 1975 on 1000 ha; in 1978 on 1900 ha and in 1981 on an area of 2666 ha (Beglyarov, 1981; Fadeev, 1981). In Poland, *P. persimilis* was used on 12 ha in 1982 and on similar areas in Roumania and Bulgaria.

In all countries in Eastern Europe, *P. persimilis* is used following regular inspections (usually every 7 days) of greenhouse crops and is released when the first spider mites are noticed. The so-called 'pest-in-first' method has not so far been practised.

In Poland, 6–10 predators are usually released per m² (=2–3 per plant). In the USSR, during the summer, up to 100 predators per m² are released in cucumber crops grown in heated glasshouses, while in spring an average of 50 predators per m² are used in plastic tunnels (Begljarov, 1978). The predator is released on infested cucumber plants by placing among them leaves of soya bean on which the predator is mass-reared.

One to three releases are as effective as 28–32 chemical treatments. The yield of cucumber plants protected by biological means increased by 2.4–5.4 kg/m^2. The profit from biological protection is very high – 1 ruble spent for this purpose gives a profit of 10–30 rubles (Begljarov, 1978).

Analysis of profits following biological control of spider mites in Poland was made by Pruszynski (1982).

Entomophthora SPP. Some species of the genus *Entomophthora* are highly infective to *Tetranychus urticae* and *T. cinnabarinus.* Zilberminc *et al.* (1978) demonstrated this in the case of *Entomophthora adjarica* isolated from a natural population of *T. urticae* in which it caused a heavy epizootic.

Egina & Cinovskij (1980) observed 73.6% mortality of *T. urticae* sprayed with conidia of *E. thaxteriana.*

1.2b **WHITEFLY CONTROL** Whitefly (*Trialeurodes vaporariorum*) is considered to be one of the most difficult pests to control in Eastern Europe. While, in the USSR, special attention is given to the use of pathogenic fungi, *Aschersonia* spp., in other countries, studies concentrate on the use of a parasitic hymenopteran, *Encarsia formosa*, and a predatory bug, *Macrolophus costalis.*

Encarsia formosa This parasite was introduced to the USSR and Poland from Canada in 1960, while further introductions into Poland were made from the USSR in 1967 (Kowalska, 1969). It has also been introduced to Bulgaria, Czeckoslovakia, the German Democratic Republic, Hungary and Roumania, but so far has not had such wide use as *P. persimilis.* In 1982, *E. formosa* was used on the following areas of tomato production: Bulgaria 5 ha, Poland 1.5 ha, USSR 3 ha.

As soon as the first plants infested with whiteflies are noticed, the release of *E. formosa*, by hanging up leaves with parasitized whitefly pupae, is recommended. Ten *E. formosa* per m^2 are released in 2 or 3 doses at intervals of 2 weeks. This provides 95% control of whitefly. As shown by Popov & Zabudskaja (1983), 1 ruble spent on biological control of whitefly gave a profit of 5.39 rubles. In greenhouses where *E. formosa* was used, the tomato yield was 23.47 kg/m^2 but where plants were protected only with chemicals the yield was 16.23 kg/m^2.

Macrolophus costalis This predatory bug belonging to the family Miridae, shows some promise for control of *T. vaporariorum* and is being studied in Bulgaria, the USSR and Poland (Brzesinski, 1982).

Aschersonia spp. A number of species of this fungus were introduced from China to the USSR in 1958 for biological control of *Dialeurodes citri* on citrus trees. Izevskij & Prilepskaja (1978) described in detail the history of *Aschersonia* use for whitefly control. The first experimental trials of *A. placenta* f. *vietnamica* B. et Br. against *T. vaporariorum* were done on Sachalin in 1971 but without success (10% control). However, in the Smolensk region and in the Ukraine, using *A. placenta* and *A. flava*, good control (70–97.9%) of whitefly was obtained on cucumbers. In different years up to 10 ha of cucumbers were protected against whitefly with *Aschersonia* spp. in the USSR. The production was achieved in bio-laboratories on state farms which created difficulties in the preservation of high standards. One 0.5 l culture bottle supplies enough spores to prepare 220–250 l of water suspension containing 10^6 spores which is sufficient to spray 1000 m^2.

On a limited scale, *Aschersonia* spp. was used to control whitefly in Poland and Bulgaria.

1.2c **APHID CONTROL** In greenhouses with effective prophylactic measures, aphids do not occur on the plants before May. However, in hot and dry weather, the first infestations may occur earlier. Chemical control requires 5–6 treatments at intervals of 8–10 days during the summer.

Search for effective predators of aphids has been very intensive in many Eastern European countries and resulted in many promising developments.

Aphidoletes aphidimyza Rond. Zorin (1932, 1934 see Beglyarov, 1981) made the first attempts to use *A. aphidimyza* to control *Aphis gossypii* on cucumber in the USSR. At present, this predator is studied and used practically in the USSR, Poland and Czechoslovakia.

Bondarenko (1978) pointed out that two men can rear more than 800 000 cocoons of *A. aphidimyza* (40 000–50 000 per week) using the aphid, *Megoura viciae*, as food. According to Beglyarov (1981), *A. aphidimyza* is used against *A. gossypii* in the USSR every year on an area of about 40 ha of cucumber.

One larva kills 22–27 *A. gossypii* during its development. It is recommended that cocoons of *A. aphidimyza* are introduced 3 times at intervals of 5 days at a ratio of 1:25.

Although larvae of *A. aphidimyza* are less voracious than chrysopids or syrphids, they are more effective as they do not migrate away from aphid colonies. Furthermore, *A. aphidimyza* is effective at realtively low aphid densities. It could, therefore, be used together with chrysopid larvae which are more voracious but do not provide a permanent control.

Chrysopa carnea and *C. septempunctata* Begljarov & Yscekov (1972) first studied the possibility of using chrysopids to control *A. gossypii* and other aphids. They worked out mass-rearing methods for *C. carnea* and *C. septempunctata* as well as a method of larval release. The economic threshold for *A. gossypii* on cucumbers was found to be 1000 aphids per plant. During its development, one larva of *C. carnea* eats 200–300 *A. gossypii* or *Myzus persicae.*

Unfortunately, neither chrysopid species established or multiplied in the greenhouses. Therefore, they can be used only as a 'live insecticide' with 4–9 releases of second instar larvae during the season to effectively protect plants. Best control of *A. gossypii* on cucumber occurred when 1:1 to 1:10 larvae of *C. carnea* or 1:20 *C. septempunctata* were released.

In the case of *Aulacorthum solani* control on lettuce, the ratio of *C. carnea* larvae should be 1:30, compared with *A. gossypii* on celery at 1:25, *M. persicae* on leaf beet at 1:10 and on carnation at 1:100.

Two releases of second instar larvae at a ratio of 1:25 on celery and 1:50 on lettuce at 2-week intervals protected these plants against *M. persicae* for the whole season.

Larvae of *C. septempunctata* are more effective than *C. carnea* and, for effective colonization of the former, less are required. However, mass rearing of this species is more difficult and laborious than for *C. carnea.*

Leis axyridis This coccinellid was introduced from the Far East region of the USSR to western areas and showed some promise against *Aphis gossypii* and *Myzus persicae* in greenhouses (Sidljarevic & Voronin, 1973). Larvae of *L. axyridis* are very voracious, especially at temperatures of 22–30°C, and each kills 200–300 aphids during its life. A similar number is also killed by the adult coccinellid. Releases of larvae on cucumbers at a ratio between 1:10 and 1:30 rapidly reduced the number of *A. gossypii* over 8–9 days but did not eliminate the pest. The adults show a strong migratory tendency and are, therefore, not as effective as larvae.

Diaeretiella rapae This parasitic hymenopteran was introduced to the USSR from France in 1968. It was shown that *D. rapae* can effectively protect aubergines against *Myzus persicae.* When this parasite was released at a ratio between 1:10 and 1:50, the aphid population was destroyed in 44 days and, at the ratio of 1:100, in 55 days.

1.2d ***THRIPS* CONTROL** So far, *T. tabaci* is not considered to be a serious pest of greenhouse plants and studies on its biological control are limited. According to Begljarov (1978), P.V. Zorin, between 1932–34, introduced a predatory thrips, *Scalothrips acariphagus* Julch., from Middle Asia to some greenhouses near Leningrad to control thrips (evidently *T.*

tabaci) on cucumbers. However, this failed and the attempt was not repeated.

During 1981–82, a predatory mite, *Amblyseius mackenziei*, was introduced from the Netherlands to Poland and the USSR to control *T. tabaci* (Pruszynski, personal communication).

Bondarenko & Moiseev (1979) sprayed cucumber plants with 0.25% water suspension of a commercial fungal insecticide Boverin® containing, 6 million million spores/g of *Beauveria bassiana* (Balss.). Two applications at 10-day intervals reduced *T. tabaci* populations by 82.7% while, on plants sprayed with malathion EC 50%, at a concentration of 0.3%, the pest population was reduced by only 11.2%.

1.2e **BIOLOGICAL CONTROL OF PLANT PATHOGENS** Control of greenhouse plant pathogens by biological means receives much attention in the USSR (Fedorincik, 1978).

A commercial antibiotic, Trichotecin®, WP 10% is available to control powdery mildew on cucumbers. Trichotecin contains an antibiotic produced by *Trichotecium roseum* and has an oral toxicity to mice at the level $LD_{50} = 63–400$ mg/kg.

Trichotecin® WP 10% is recommended for use at 2 kg/ha using 1000 l of water suspension per 1 ha. The first treatment should be made when the first symptoms of mildew are noticed and later treatments repeated at 7–8 day intervals depending on the development of disease.

This fungicide is non-toxic to *Phytoseiulus persimilis*. It causes, however, some allergenic effects on man, affecting the mucous tissues. The harvest interval is 3 days, and the tolerance level is 1 mg/kg in cucumbers and 0.2 mg/m³ in air.

In many state and co-operative farms, a biological fungicide, Trichodermin-3®, is produced and used against *Rhizoctonia solani*, *Colletotrichum* spp., *Fusarium* spp. and other fungi attacking the roots and lower stems of both cucumber and tomato plants. This preparation is based on a fungus, *Trichoderma lignorum*, and is produced by growing mycelia on sterilized turf at a temperature of 24–28°C. In 1978, Trichodermin® was used on an area of 190 ha.

It is expected that another antibiotic, Arenarin®, recommended for control of tomato bacterial canker (*Bacterium michiganense*), will successfully pass the registration process. Arenarin® is used at a concentration of 1:1000 for soaking tomato seeds (100 ml of Arenarin® + 4.9 l of water + 1 kg tomato seeds) for 2.5 hours. The seeds are then dried and should be sown within 2 months. Besides its antibacterial effect, Arenarin® stimulates the growth of plants: seedlings appearing 1–6 days earlier, flowering 2–3 days earlier, first fruits – 3–5 days earlier. Yield increases vary from 5–40%.

1.2f **BIOLOGICAL CONTROL OF RODENTS** Biological control of mice and related rodents in greenhouses and hotbeds has been used in the USSR for several years. Two commercial microbial rodenticides, both containing cells of *Salmonella enteritidis* var. *Issatschenko*, are available.

Grain Bactorodencide® is recommended for use against house mouse, field mouse, common vole, harvest mouse and others. Each grain contains 150–300 million bacterial cells and 2–3 grains are lethal to mice.

Bactorodencide® is especially effective against rats and house mice and is produced on animal peptides and bones. It is mixed with baits (flour, potato etc.) and some water before use. In 1978, both types were used on 550 ha of greenhouse and hotbeds.

Obviously, a wide range of natural methods of control are practised in Eastern Europe but there is little evidence that integrated programmes embracing all aspects of pest and disease are practised. It should also be noticed that their use of antagonists and rodenticides is facilitated by less stringent registration requirements.

REFERENCES

Begljarov, G.A. (1978) (Biological control of pests of vegetables in greenhouses) pp. 52–62. In *Zasozta Ovoscnych, Bachcevych Kultur i Kartofelja ot Vreditelej, Boleznej i Sornjakov* Kolos, Moskva. 168 pp. (In Russian.)

Begljarov, G.A. (1981) Advances in and outlook for development of biological control to protect plants under glass in the USSR. pp. 36–37. In *Proceedings of the Joint American–Soviet Conference on Use of Beneficial Organisms in the Control of Crop Pests* (Coulson, J.R., ed.) Entomological Society of America. 62 pp.

Begljarov, G.A., Vasilev, R.A., Chlopceva, R.I. & Listkova, R.A. (1964) (Developing biological methods of controlling spider mites in greenhouses.) pp. 119–122. In *Issleddvanija po Biologiceskomy Metodu s Vrediteljami Selskogo i Lesnogo Chozjajstwa. Doklady k simpozjum* October 1964. Novosibirsk. 224 pp. (In Russian.)

Begljarov, G.A. & Yscekov, A.T. (1972) (On possibilities of use of common aphid-lion to control aphids in glasshouses.) pp. 33–43. In *Biologiceskij Metod Borby s Vrediteljami Ovoscnych Kultur* Kolos, Moskva. 112 pp. (In Russian.)

Beglyarov, G.A. see Begljarov, G.A.

Bondarenko, N.V. (1978) (Use of predatory cecidomyids to control aphids in greenhouses.) pp. 22–23. In *Biologiceskij Metod Borby s Vrediteljami i Boleznjami Rastenij v Zakrytom Grunte* (Begljarov, G.A. & Cekmenev, S. Ju., eds) Kolos, Moskva. 176 pp. (In Russian.)

Bondarenko, N.V. & Moiseev, E.G. (1979) (Biological protection of cucumber against tobacco thrips in greenhouses.) *Zashch. Rast. Vredit. Bolez.* **379**: 3–50.

Brzesinski, K. (1982) (Report from investigations on morphology, biology and ecology of the heteropteran *Macrolophus costalis* (Fieb.) (Heteroptera, Miridae) and its predacity in relation to greenhouse whitefly (*Trialeurodes vaporarium* Westw.).) pp. 283–292. In *Materialy XXII i XXIII Sesji Naukowe IOB Poznan* 319 pp.

Egina, K. Ja. & Cinovskij, Ja. P. (1980) (Results of evaluation of efficacy of a fungus *Entomophthora thaxteriana* (Petch) Hall et Bell on aphids and spider mites after storage.) pp. 7–11. In *Biologiceskij Metod Borby v Vrednymi Nasekomymi i Klescami* (Cinovskij, Ja. P., ed.) Zinatne, Riga. 73 pp.

Fadeev, Y.N. (1981) Prospects for development of biological methods in plant protection in the USSR. pp. 12–13. In *Proceedings of the Joint American–Soviet Conference on Use of Beneficial Organisms in the Control of Crop Pests* (Coulson, J.R., ed.) Entomological Society of America. 62 pp.

Fedorinck, N.W. (1978) (Use of micro-organisms and products of their metabolites for biological control of plant diseases in greenhouses.) pp. 151–161. In *Biologiceskij Metod Borby s Vrediteljami Boleznjami Rastenij v Zakrytom Grunte* (Begljarov, G.A. & Cekmenev, S.Ju., eds) Kolos, Moskva. 176 pp. (In Russian.)

Izevskij, S.S. & Prilepskaja, N.A. (1978) (Application of fungi of the genus *Aschersonia* in controlling whitefly.) pp. 115–124. In *Biologiceskij Metod Borby s Vrediteljami i Boleznjami Rastenij v Zakrytom Grunte* (Begljarov, G.A. & Cekmenev, S.Ju., eds) Kolos, Moskva. 176 pp. (In Russian.)

Kandybin, N.V. (1978) (Microbiological means of plant protection against rodents and insects in hotbeds and greenhouses.) pp. 99–115. *Biologiceskij Metod Borby s Vrediteljami i Boleznjami Rastenij v Zakrytom Grunte* (Begljarov, G.A. & Cekmenev, S.Ju., eds) Kolos, Moskva. 176 pp. (In Russian.)

Kowalska, T. (1969) (Introduction of *Encarsia formosa* Gah., a parasite of whitefly (*Trialeurodes vaporarium* Westw.).) *Biul. Inst. Ochr. Rosl.* **44**: 341–551.

Kowalska, T., Szczepanska, K. & Bartkowiak, A. (1974) (Trials on biological control of aphids in greenhouses.) *Biul. Inst. Ochr. Rosl.* **57**: 281–289.

Lipa, J.J., Pruszynski, S. & Wgorek, W. (1967) (Results of preliminary studies on acclimatization of *Phytoseiulus persimilis* Athias-Henriot against spider mites in Poland.) *Biul. Inst. Ochr. Rosl.* **36**: 87–92. (In Polish.)

Popov, N.A. & Zabudskaja, I.A. (1983) (Use of *Encarsia* on cucumbers.) *Zashch. Rast. Vredit, Bolez.* **59**) No. 3: 2 pp.

Pruszynski, S. (1982) (Biological and integrated methods of protection of glasshouse cultures against pests – advantages and perspectives.) pp. 293–301 in *Materialy XXII i XXIII Sesji Naukowe IOB Poznan*, 319 pp.

Sidljarevic, V.J. & Voronin, K.E. (1973) (An experiment of use of harmonia in greenhouses.) *Zashch. Rast. Vredit. Bolez.* **49**, No. 6: 24 pp.

Zilberminc, I.V., Cindadze, K.V. & Vartapetev, S.G. (1978) (Entomopathogenic fungus *Entomophthora adjarica* Cinz. Vart., perspective for use to control spider mite in greenhouses.) pp. 125–134. In *Biologiceskij Metod Borby s Vrediteljami i Boleznjami Rastenij v Zakrytom Grunte* (Begljarov, G.A. & Cekmenev, S.Ju., eds) Kolos, Moskva. 176 pp. (In Russian.)

2 BIOLOGY OF PESTS AND NATURAL ENEMIES

2.1 WHITEFLY AND ITS PARASITE *ENCARSIA FORMOSA*

C. Stenseth

Vet *et al.* (1980) have provided a general review of the literature on the biology of both the greenhouse whitefly and *Encarsia formosa.*

2.1a **GREENHOUSE WHITEFLY (*TRIALEURODES VAPORARIORUM*)**

249 genera are known to be host plants of the greenhouse whitefly, *Trialeurodes vaporariorum* (Plate 1). Among the greenhouse host plants are cucumber, aubergine, paprika, tomato and many ornamentals. Among the latter are *Azalea, Calceolaria, Fuchsia, Pelargonium,* poinsettia and *Verbena.*

The adult whitefly is attracted to yellow colours but is unable to detect whether or not a plant is a suitable host before alighting. The females lay their eggs on the undersides of young apical leaves, often in circles on hairless hosts. The eggs hatch after 8 days (21–24°C) while further development consists of 1st, 2nd, 3rd and 4th instars and pupae whose development, at these temperatures, occupies 6, 2, 3, 4 and 5 days respectively. The newly hatched larvae ('crawlers' – 1st instar) are initially mobile. They move for a few hours only and then settle. After inserting their mouthparts into leaf tissue, they lose their functional legs and remain static throughout the remainder of their development. After the third moult, there are three additional phases of development. During the first (4th instar), the larvae become flattened and in the second, this instar becomes thickened with lateral spines. The third stage, the pupa, is of similar shape to the second stage, but the red eyes of the developing adult become visible.

Temperature governs the rate of growth so that the total developmental time varies from 18–123 days (Table 1). Mortality in the different developmental stages varies with the type of host plant (Table 2).

The adults emerge through a slit in the dorsum of the pupa. The females commence oviposition after 1–2 days. Unmated females lay haploid eggs which produce males. Mated females produce both diploid female-producing eggs and haploid eggs. The longevity of the adult females depends on the host plant (Table 2). Those host plants encouraging greatest longevity are also the most suitable for oviposition

(Table 2). Egg production increases with increasing density of whiteflies per leaf; it is also greater at high (26°C) rather than moderate temperatures (21°C), but the highest total egg production occurs at around 21°C. A humidity of 75–80% RH is optimal for both fecundity and longevity of the adults.

The greenhouse whitefly has no stage specially adapted for hibernation. Survival is dependent on suitable host plants throughout the year but, at low temperatures, the plant must have winter-hardy leaves. Eggs are the stage most tolerant to low temperatures and can survive up to 15 days at −3°C but only 5 days at −6°C.

TABLE 1 Duration of life cycle in *Trialeurodes vaporariorum* and *Encarsia formosa* from egg to adult (in days) at different temperatures (Stenseth, 1971 & 1975)

Species	12°C	15°C	18°C	21°C	24°C	27°C	30°C	Host Plant
T. vaporariorum	103–123	65–72	37–42	25–30	22–25	—	18–21	Bean
E. formosa	—	—	29–39	25–35	16–24	13–17	—	Tomato

TABLE 2 The effect of the host plant on longevity in adult females of *Trialeurodes vaporariorum*, their egg production and mortality in the different developmental stages (after Woets & van Lenteren, 1976)

	Aubergine	Cucumber	Tomato	Sweet pepper
Longevity (days)	40.4	16.7	8.6	3.2
Total number of eggs	416	123	8.2	0.9
Mortality (%)	9.2	7.4	21.7	92.5

2.1b THE PARASITE *ENCARSIA FORMOSA* The developmental stages of this parasite consist of an egg, three larval stages, a pupa and the adult (Plate 2). With the exception of the adult, development takes place within its host – the larvae and pupa of greenhouse whitefly.

The adult parasite is attracted to its host by a volatile compound emanating from the whitefly honeydew which can be detected over several metres (Ledieu, 1976). It usually lands to search for hosts only on whitefly-infested plants and seldom on uninfested plants. The adult feeds both on whitefly honeydew and on body fluids through a hole made in the whitefly larvae with the ovipositor. Males are rare and result from oviposition in parasitized scales (hyperparasitism) which may occur when the density of parasites is high. Cool temperatures also appear to increase production of males. Normally the sex ratio is 1:1 but twice as many as females may develop.

The wasp can distinguish between unparasitized and parasitized hosts, the latter being avoided for oviposition unless parasite density is high. Oviposition can take place in all four larval instars and the pupa but the parasites prefer the 3rd and 4th instars. The least mortality occurs in these stages and consequently these instars provide the best chances of successful parasitism. Oviposition in the 1st and 2nd whitefly instars results in high mortality of the parasite; consequently many of the young scales shrivel and die. Close examination will reveal multiple oviposition punctures. This mortality may, in some circumstances, exceed that of parasitism.

Under laboratory conditions, the females have a longevity of about 27, 21, 15, 8 and 3 days at temperatures of 18°, 21°, 24°, 27° and 30°C respectively. In a greenhouse experiment, the number of introduced parasites active on the plant was reduced to half in 4 days at 18°C and at 24–27°C a longevity of only 2–3 days was observed. Low light intensity also reduces the longevity of adult parasites.

The mean egg production is commonly reported to vary from 50–100 eggs per female, but as many as 350 eggs have been recorded from single females when single plants are exposed in the laboratory. Between 18 and 27°C there are only small differences in parasite fecundity. Most parasitization is recorded at humidities of 50–80% RH. Low light intensities seem to have a strong influence on fecundity. Very few eggs were laid below 4200 lux but at 7300 lux the parasites became fully reproductive in laboratory experiments at GCRI. Experience in UK greenhouses also shows that introduction of the parasite during the winter (before 1 March) achieves too low a parasitization for practical use. The physical structure of the host plant also influences parasitism. It is known that, on cucumber leaves, *E. formosa* attacks fewer hosts than on tomato or aubergine. It is assumed that this effect is caused by the hairy leaves of cucumber, which reduce the walking speed of the wasp and contaminate it with honeydew from the glandular hairs. The wasp, therefore, takes longer to clean its body on cucumber than on plant hosts with a different hair structure. Another important factor is that, independent of the host plant, parasitization declines as the quantity of whitefly honeydew increases (Parr, 1969). It follows, therefore, that dense whitefly populations hamper the parasite.

A major factor affecting parasite efficiency is temperature in that flight is said to be inhibited below 17°C. However, van Lenteren & Hulpas (1983) have recently demonstrated flight at 12°C but there seems little doubt that direct radiation from sunlight plays an important role.

The duration of the parasite life cycle depends on the temperature (Table 1). When half the development of the parasite is completed, the whitefly host turns black and the parasite finally emerges through a hole in the dorsum of the 'black scale'.

The ecological information about greenhouse whitefly and its parasite suggests that the nature of the host plant is very important for the practical utilization of *E. formosa.* The effects of temperature and humidity are not fully understood but temperatures above 18°C and a humidity of between 50 and 80% RH seem to ensure effective parasitism and successful control of whitefly, providing the light intensity is sufficient and the whitefly population is low.

REFERENCES

Ledieu, M.S. (1976) Dispersal of the parasite *Encarsia formosa* as influenced by its host *Trialeurodes vaporariorum. Bull. IOBC/WPRS Working Group on Integrated Control in Glasshouses, Antibes* **1976** (4): 122–124.

Parr, W.J. (1969) Glasshouse whitefly, *Rep. Glasshouse Crops Res. Inst.* **1968**: 90.

Stenseth, C. (1971) Effect of temperature on the development of *Trialeurodes vaporariorum* Westwood. *Forsk. Fors. Landbr.* **21**: 357–366.

Stenseth, C. (1975) Effect of temperature on the development of the parasite *Encarsia formosa. Gartneryrket* **65**: 136–139.

van Lenteren, J.C. & Hulpas-Jordaan, P.M. (1983) Influence of low temperature regimes on the capability of *Encarsia formosa* and other parasites in controlling greenhouse whitefly. *Bull. IOBC/WPRS Working Group on Integrated Control in Glasshouses, Darmstadt* **6** (3): 54–70.

van Lenteren, J.C. & van der Scheel, A. (1981) Temperature thresholds for oviposition of *Encarsia formosa, E. tricolor* and *E. partenopea. Meded. Rijksfac. Landb. Gent* **46**: 456–483.

Vet, L.E.M., van Lenteren, J.C. & Woets, J. (1980) The parasite-host relationship between *Encarsia formosa* (Hymenoptera: Aphelinidae) and *Trialeurodes vaporariorum* (Homoptera: Aleyrodidae). *Z. Angew. Ent.* **90**: 26–51.

Woets, J. & van Lenteren, J.C. (1976) The parasite-host relationship between *Encarsia formosa* (Hymenoptera: Aphelinidae) and *Trialeurodes vaporariorum* (Homoptera: Aleyrodidae). *Bull. IOBC/WPRS Working Group on Integrated Control in Glasshouses, Antibes* **1976** (4): 151–164.

2.2 BIOLOGY OF THE FUNGI *VERTICILLIUM* AND *ASCHERSONIA*

R.A. Samson & M.C. Rombach

Several entomopathogenic fungi have been studied in connection with the biological control of insect pests and recently some success has been achieved (Roberts & Humber, 1981; Burges, 1981; McCoy *et al.*, 1984). Examples are *Hirsutella thompsonii* against the citrus rust mite, *Nomuraea rileyi* against lepidopterous larvae and *Metarhizium anisopliae* against spittle bugs and *Rhinoceros* beetle. In glasshouse crops, research has focussed on two fungi for use as microbial pesticides: *Verticillium lecanii* for control of aphids and whitefly, and *Aschersonia aleyrodis* for whitefly control. The latter has aroused interest in Western Europe only recently, despite successful experiments elsewhere. However, recent Dutch research (Ramakers & Samson, 1983) has given promising results. The development of *V. lecanii* as a microbial insecticide has progressed well because of studies by researchers at GCRI (see Hall, 1981).

These two fungi differ in their biology and ecology and therefore in their production and application techniques. These differences are discussed below.

2.2a ***VERTICILLIUM LECANII*** *V. lecanii* was first reported in 1939 by Viegas, who referred to the characteristic white halo formed by the fungus on the scale insect *Coccus viridis* (Green) as 'the farmers' friend'. Recently the effectiveness on this insect of *V. lecanii* was again demonstrated in India by Easwaramoorthy & Jayaraj (1978).

The fungus is a well known cosmopolitan species described under several names, such as *Cephalosporium lecanii* and *Cephalosporium aphidicola*. The taxonomy of the species has been revised by Gams (1971), who placed it in the genus *Verticillium*, mainly because of the arrangement of its conidiogenous cells in regular whorls. In his taxonomic study, Gams (1971) applied a rather broad species concept of *V. lecanii* and included in the taxon both small- and large-spored strains. Recently Evans & Samson (1982), in their studies on entomogenous fungi from the Galapagos Islands, found that *V. lecanii* parasitizing coccids is associated with the ascomycete, *Torrubiella confragosa*.

V. lecanii is not restricted to insect hosts. The species is commonly isolated from mouldy organic material, foodstuffs and soil (Domsch *et al.*,

FIGURE 7 Stereoscan of healthy *Myzus persicae*.

FIGURE 8 Stereoscan of *Myzus persicae* attacked by fungus, *Verticillium lecanii*.

1980; Samson *et al.*, 1980). It is hyperparasitic on various fungi, such as rusts, agarics and even entomogenous fungi, and it is facultatively parasitic on various insects and arachnids, but has not been observed as a pathogen of mammals.

On the insect, *V. lecanii* is found as cottony whitish yellow colonies (Figures 7 & 8). On the aerial mycelium are conidiophores bearing awl-shaped phialides arranged in a characteristic verticillate manner. The cylindrical to ellipsoidal conidia are aggregated in a mucus. Conidia are readily produced by *V. lecanii*, while, in submerged cultures, blastospores are formed by a yeast-like budding process. Hall (1980a) reported that virulence was not effected by successive transfers on artificial media nor by passage of conidia through the original host. Hall (1980b) isolated, from several sources, strains with different virulence towards the host, *Macrosiphoniella sanborni*. Currently, a strain with small spores for control of aphids and another with relatively large spores for control of whitefly larvae are used in commercial production. A product containing conidia of a different strain to be used against the cosmopolitan greenhouse pest, *Thrips tabaci*, is under development.

Natural infections of insect populations by *V. lecanii* are common in the glasshouses in Western Europe. The inoculum which starts these epidemics probably originates from soil or mouldy organic material. These infections can be induced and strongly enhanced by agricultural practices, e.g. the use of polyethylene blackout sheets for manipulating flowering of chrysanthemums or the frequent use of overhead spray installations, in Southern Europe.

V. lecanii grows and multiplies at temperatures of between 15 and 25°C and humidities of 85–90% RH in the greenhouse. High humidity must be present for at least 10–12 hours per day. Under these conditions, epidemics can occur within aphid or whitefly populations (see also Hall & Burges, 1979, for more detailed information on the effects of temperature and humidity). The first infected insects are observed as white cottony particles ('fluffy-bodies') adhering to the undersides of the leaves 6–12 days after spraying (Figure 9). Under favourable conditions, aphid and whitefly populations can be suppressed for several months following a single application of the fungus. Most plant pathogenic fungi also grow best at the same temperature and humidity range, e.g. *Botrytis* sp. in tomatoes.

Successful epizootic development of *V. lecanii* has often been arrested by applications of fungicides. In these cases, a second introduction of the fungus is necessary, after a safety interval of 1–2 weeks, depending on the fungicide used. Therefore, careful planning, particularly of aerial applications of fungicides, should be an important part of each integrated control programme involving the use of fungi. *V. lecanii* seems also to be sensitive to some insecticides (Section 5). However, the application of a

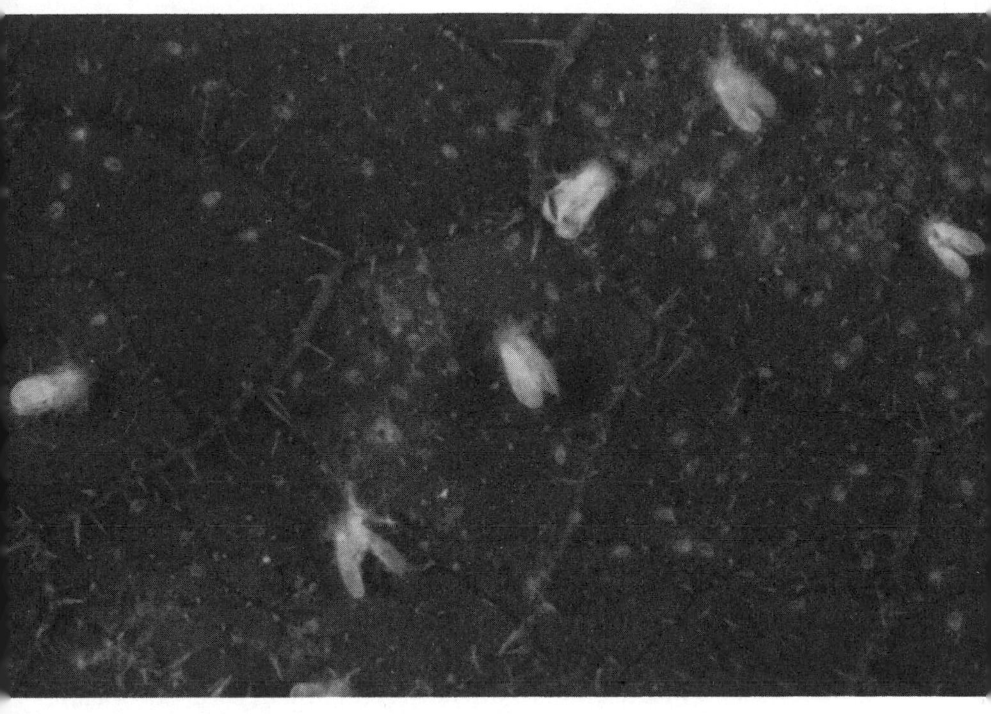

FIGURE 9 *Verticillium* infection

combination of the fungus with a low dosage insecticide (e.g. fenthion) increases mortality in populations of *Coccus viridis* (Easwaramoorthy *et al.*, 1978).

The nature of the infection process by *V. lecanii* is not known. The fungus germinates and initially grows in an apparently saprobic manner outside the insect host. This growth might occur on the honeydew excreted by the insects or, when applied as a commercial product, on the carbohydrate carrier material included in the product. Following the initial saprobic growth, infection of the insect can occur by conidia or by hyphae growing through natural orifices and between body segments. In contrast to other entomopathogenic fungi, a direct contact between a *V. lecanii* conidium and its future host is not necessary to accomplish infection.

The dispersal of conidia in the greenhouse by air movement is unlikely. The conidia are sometimes dispersed by live insects and mites, as reported for *Metarhizium anisopliae* by Schabel (1982). In this respect, predatory mites and parasitic wasps might play a significant role. *V. lecanii* can occasionally be found infecting adult *Encarsia formosa* as well (Ekbom, 1979). However, the influence of such fungal infections on the wasp population is very limited.

2.2b ***ASCHERSONIA ALEYRODIS*** The genus *Aschersonia* belongs to the family Coelomycetes of the class Fungi Imperfecti and contains over 30 species mostly found in the (sub-) tropics (Petch, 1921; Mains, 1959). All known species are highly host specific and all have been described from scale insects (Coccidae) and whiteflies (Aleyrodidae). The species which attack whiteflies parasitize only the nymphal stages and not the adult. In nature, some species are associated with species of the ascomycete, *Hypocrella*.

A. aleyrodis was first described from citrus whitefly larvae in Florida. The species resembles *A. placenta*, a common fungus on whiteflies in subtropical and tropical countries in Asia, differing mainly in the bright orange-red colour (Plate 3). After the turn of the century, *A. aleyrodis* was successfully introduced in Florida as a bio-insecticide in populations of *Dialeurodes citri* (Ashmead) by Fawcett (1908) and Berger (1907). In some regions, whiteflies are still effectively controlled by epizootics following these original applications. Recently, the species has been tested against *T. vaporariorum* in the USSR (Primak & Chiznik, 1975; Osokina & Izhevskii, 1975; Kogan & Seryapin, 1978) and in Bulgaria by Kristova (1971).

Microscopical studies on the citrus whitefly (*Dialeurodes citri* (Ashmead)) naturally infected with the fungus, *Aschersonia aleyrodis* Webber, show that the conidia are one-celled, fusiform, smooth-walled propagules produced in slime by phialides. These phialides are arranged in cavities, or pycnidia (Samson & McCoy, 1983). Sporulation occurs early in the infection process; soon after the hyphae rupture the dorsal cuticle and produce mat-like pustules of white mycelia on the host surface (Figures 10 & 11).

Infection of whitefly larvae by *A. aleyrodis* is easily detected by the naked eye because of the bright orange, slimy spore masses formed under humid conditions on the bodies of the larvae (Plate 4). In drier conditions, a yellowish discolouration of the larvae or a whitish mycelium protruding from the host can be observed.

In greenhouses, the species has been found to be highly host-specific. During several years of experimentation in the Netherlands, the species was never found infecting whitefly adults nor any other arthropod species.

The species can be successfully grown on various mycological media (Samson *et al.*, In preparation). Malt extract agar is the most favourable for isolation and maintenance in the laboratory. Deterioration of the colonies has been observed on several media currently in use (e.g. Yeast-

FIGURE 10 Sporulating structures of *Aschersonia aleyrodis* (× 3000).

FIGURE 11 Whitefly larva parasitized by *Aschersonia aleyrodis*. Early stage of infection (× 95).

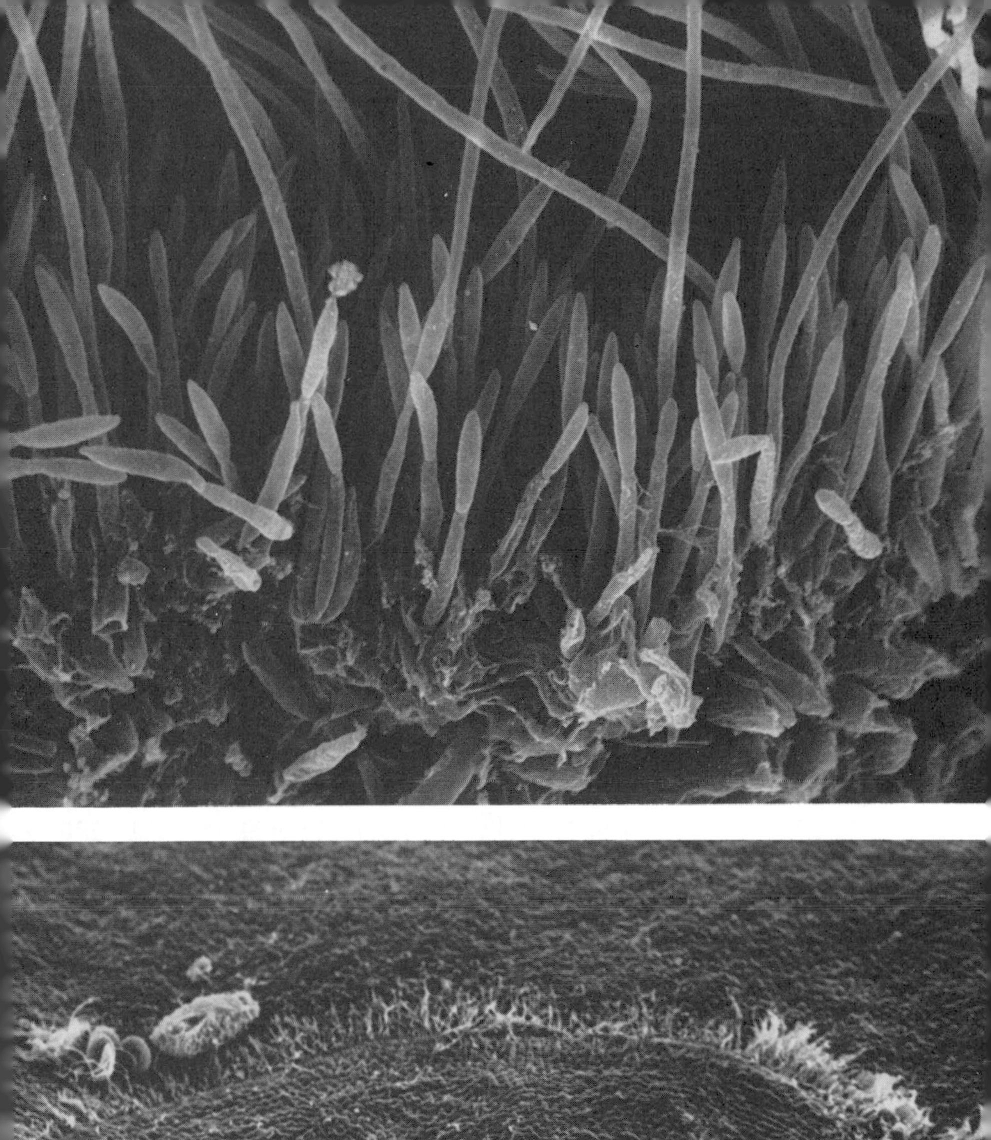

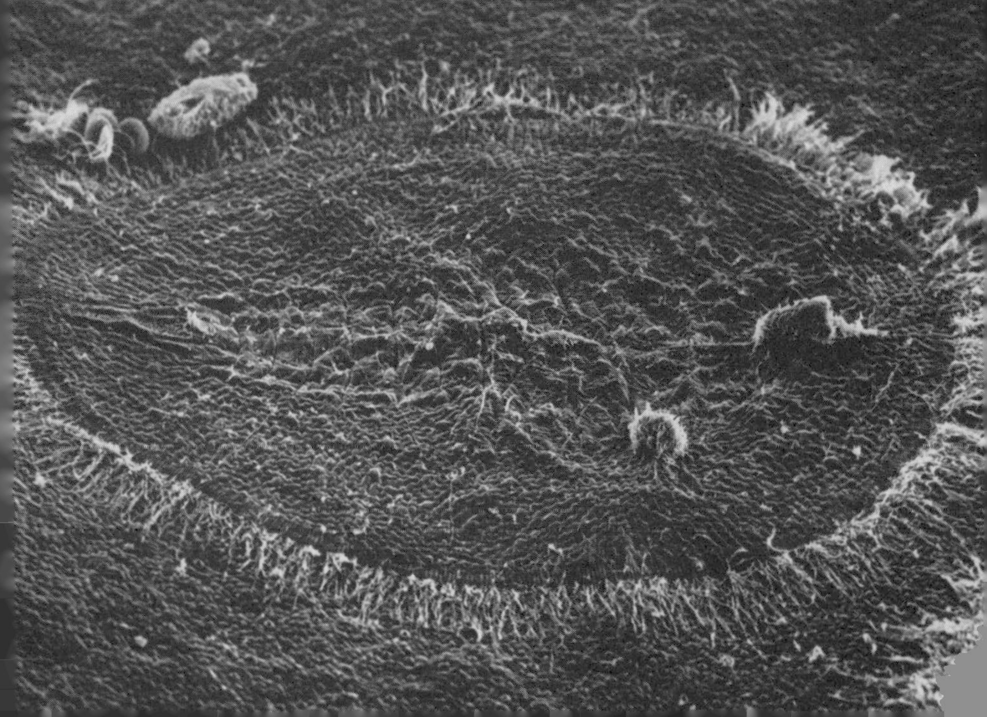

Peptone-Glucose) prior to or during sporulation. The colonies appear white and cottony after 1 week of growth. Light (350–420 nm) seems to induce sporulation and the characteristic yellow-orange pigmentation. After several transfers, the fungus grows and sporulates less vigorously; this contrasts with *V. lecanii.* This deterioration of a strain seems to be a common phenomenon in entomogenous hyphomycetes and has also been reported for the common *Metarhizium anisopliae* and *Beauveria bassiana.* Therefore, regular passage through the insect host followed by subsequent re-isolation, or suitable storage of the original isolate is recommended in order to maintain viability and infectivity.

After initial contact, conidia of *A. aleyrodis* rapidly penetrate the cuticle of whitefly larvae. The exact nature of the infection process is not yet known. Under greenhouse conditions, most infection occurs during the night following the introduction of conidia. Once inside the host, the fungus is lethal, despite low humidity conditions in the greenhouse. Therefore, closing the ventilators of the greenhouse or the sides of plastic houses for 1 night following application of the fungus is sufficient to ensure infection. The first macroscopical symptoms of infection appear after 8 days. Melting of the two large fat bodies within whitefly larvae may be observed (Landa, Personal communication, 1979).

The absence of an initial phase of saprobic growth makes necessary direct contact between a conidium and its future host. Therefore, emphasis has to be placed on adequate spray techniques when using preparations of this fungus. Clustering of the conidia in the spray solution can be avoided by the addition of detergents.

Control of the greenhouse whitefly (*Trialeurodes vaporariorum*) by treating young larvae with conidia of *A. aleyrodis* has been studied in heated greenhouses in the Netherlands. Artificial infection succeeded both in rainy periods and during sunny weather, but spontaneous reinfection was not observed. A dose of 2×10^8 conidia of *A. aleyrodis*/plant, applied as an ultra low volume spray to a cucumber crop, caused 75% mortality. *A. aleyrodis* never occurred in the control plots. The fungus did not affect the ratio between numbers of whitefly and its parasite, *Encarsia formosa* (Ramakers & Samson, 1983).

Because of the lack of dispersing agents (e.g. wind and rain) in the greenhouse, repeated applications are necessary to maintain the pathogen in the insect population. Kogan & Seryapin (1978) and Ramakers & Samson (1983) reported that, to accomplish almost complete kill, a dose of about 10^{13} conidia/ha was necessary in a cucumber crop. Kristova (1971) claimed permanent control by additional sprays of water following a single application of the fungus, but her findings could not be confirmed in our experiments.

Little is known about the safety of *A. aleyrodis* for human and other vertebrates. However, the fungus does not grow above 31°C and no cases are known suggesting that the fungus is toxic or pathogenic.

2.2c **CONCLUSIONS** The different characteristics of *A. aleyrodis* and *V. lecanii* open a wide spectrum of possible uses for these fungi within integrated control programmes. The choice of fungus depends upon the environmental conditions and the efficacy of the other biological control agents in use.

REFERENCES

Berger, E.W. (1907) Whitefly conditions in 1906. The use of fungi. *Bull. Fla. Univ. Agric. Exp. Stns. (Gainesville)* **88**: 1–85.

Burges, H.D., ed. (1981) *Microbial Control of Pests and Plant Diseases 1970–1980* Academic Press, London. 949 pp.

Domsch, K.H., Gams, W. & Anderson, T. (1980) *Compendium of Soil Fungi* Vols I & II. Academic Press, London. 859 + 405 pp.

Easwaramoorthy, S. & Jayaraj, S. (1978) Effectiveness of the white halo fungus *Cephalosporium lecanii* against field populations of coffee green bug, *Coccus viridis. J. Invert. Path.* **32**: 88–96.

Easwaramoorthy, S., Regupathy, A., Santharam, G. & Jayarai, S. (1978) The effect of subnormal concentrations of insecticides in combination with the fungal pathogen, *Cephalosporium lecanii* Zimm. in the control of coffee green scale, *Coccus viridis* Green. *Z. Angew. Ent.* **86**: 161–166.

Ekbom, B.S. (1979) Investigations on the potential of a parasitic fungus (*Verticillium lecanii*) for the control of greenhouse whitefly (*Trialeurodes vaporariorum*). *Swed. J. Agric. Res.* **9**: 129–138.

Evans, H.C. & Samson, R.A. (1982) Entomogenous fungi from the Galapagos Islands. *Can. J. Bot.* **60**: 2325–2333.

Fawcett, H.S. (1908) Fungi Parasitic upon *Aleurodes citri.* Ph.D. Thesis, University of Florida.

Gams, W. (1971) Cephalosporium-*artige Schimmelpilze (Hyphomycetes).* Gustav Fischer Verlag, Stuttgart.

Hall, R.A. (1980a) Effect of repeated subculturing on agar and passage through an insect host on the pathogenicity, morphology and growth rate of *Verticillium lecanii. J. Invert. Path.* **36**: 216–222.

Hall, R.A. (1980b) Laboratory infections of *Verticillium lecanii* strains isolated from phytopathogenic fungi. *Trans. Br. Myc. Soc.* **74**: 445–446.

Hall, R.A. (1981) The fungus *Verticillium lecanii* as a microbial insecticide against aphids and scales. In *Microbial Control of Pests and Plant Diseases 1970–1980* (Burges, H.D., ed.). Academic Press, London. 949 pp.

Hall, R.A. & Burges, H.D. (1979) Control of aphids in glasshouses with the fungus *Verticillium lecanii. Ann. Appl. Biol.* **93**: 235–246.

Kogan, V.Sh. & Seryapin, A.A. (1978) (Against the greenhouse whitefly). *Zashch. Rast. Vredit. Bolez.* **23**: 53–56. (In Russian.)

Kristova, E. (1971) Possibilities of using *Encarsia formosa* Gahan and *Aschersonia aleyrodis* Webber in the control of *Trialeurodes vaporariorum* Westw. *Tech. Commun. ISHS* **17** (II): 396–399.

McCoy, C.W., Samson, R.A. & Boucias, D. (1984) Entomogenous fungi. In *Handbook of Naturally Occurring Pesticides. Vol. V. Microbial Insecticides* (Ignoffo, C., ed.). CRC Press Inc. Boca Raton, USA (In press).

Mains, E.B. (1959) North American species of *Aschersonia* parasitic on *Aleyrodidae*. *J. Invert. Path.* **1**: 43–47.

Osokina, G.A. & Izhevskii, S.S. (1975) (A test on the control of the greenhouse whitefly.) *Zashch. Rast. Vredit, Bolez.* **21**: 28–29. (In Russian.)

Petch, T. (1921) Studies in entomogenous fungi II. *Hypocrella* and *Aschersonia. Ann. R. Bot. Gn. Peradeniya* **7**: 167–278.

Primak, T.A. & Chiznik, R.I. (1975) (The possibilities of using the fungus *Aschersonia aleyrodis* for the control of the greenhouse whitefly.) *Zakh. Rosl.* **22**: 53–56. (In Russian.)

Ramakers, P.M.J. & Samson, R.A. (1983) *Aschersonia aleyrodis*, a fungal pathogen of whitefly. II. Application as a biological insecticide, in glasshouses. *Z. Angew. Ent.* (In press).

Roberts, D.W. & Humber, R.A. (1982) Entomogenous fungi. In *Biology of Conidial Fungi.* Vol. 2. (Cole, G.T. & Kendrick, B., eds) Academic Press, New York & London.

Samson, R.A., Hoekstra, E.H.S. & van Oorschot, C.A.N. (1980) *Introduction to Food-borne Fungi.* CentraalBureau voor Schimmelcultures, Baarn, Netherlands.

Samson, R.A. & McCoy, C.W. (1983) *Aschersonia aleyrodis*, a fungal pathogen of whitefly. I. Scanning electron microscopy of the development on citrus whitefly. *Z. Angew. Ent.* (In press).

Samson, R.A., Rombach, M.C. & Ramakers, P.M.J. (1984) *Aschersonia aleyrodis*, a fungal pathogen of whitefly. III. Isolation and cultivation on artificial media. (In preparation).

Schabel, H. (1982) Phoretic mites as carriers of entomopathogenic fungi. *J. Invert. Path.* **39**: 410–412.

2.3 RED SPIDER MITE AND THE PREDATOR *PHYTOSEIULUS PERSIMILIS*

N.E.A. Scopes

2.3a **RED SPIDER MITE (*TETRANYCHUS URTICAE* [*T. TELARIUS*])** Red spider mite, also known as the two-spotted mite (Plate 5), is a major pest of both ornamental (Figure 12) and vegetable plants throughout the world. The extensive world literature has been well summarized by Hussey & Huffaker (1976).

T. urticae has five developmental stages: egg, larva, protonymph, deutonymph and adult, each nymphal stage having both a feeding and resting stage. Females lay spherical eggs (0.14 mm diameter) on the undersides of leaves. The small, whitish larva has three pairs of legs while the protonymph and older stages have eight legs. Sexual characters become obvious at the deutonymph stage, those individuals with elongated bodies developing into males and those with rounded bodies into females.

Males are attracted to female deutonymphal resting stages by pheromones, so that females mate immediately on maturing and begin

FIGURE 12 Red spider mite damage on azalea.

FIGURE 13 Webbing by spider mites on chrysanthemum flowers.

laying eggs within 36 hours. Unfertilized eggs produce males while fertilized ova produce both males and females. At 20°C and 36% RH a female will lay 7.3 eggs/day while at high humidities (95% RH) reproduction declines to about 4.9 eggs/day. The multiplication rate per generation is some 31 ×. The life cycle at 32, 21, 18 and 15.5°C takes 3.5, 14.5, 21 and 30 days respectively. The normal sex ratio (♀:♂) is 3:1.

Mites feed on cell chloroplasts, producing characteristic minute, yellowish, speckled feeding marks which may coalesce, causing leaves to shrivel and die. If the number of mites on a plant becomes excessive they migrate to the apical leaves where they produce silken webs (Figure 13). Individuals may then drop as much as a metre on a silken thread down which others climb creating a 'rope' with a ball of jostling individuals at the end. Such ropes are readily dispersed to other plants, either by wind or by physical transport by workers or equipment.

Because of their small size and rapid rate of multiplication, numbers are best estimated by relating them to damage symptoms (Hussey & Parr, 1963) (see Figure 14 & Table 3).

TABLE 3 Definition of intensities of mite damage to cucumber leaves (Hussey & Parr, 1963)

Damage index	Definition	Mite population (adults plus nymphs) per 6.45 cm^2
0	No damage	0
1	Incipient damage, on or two 120 mm feeding patches	3
2	Feeding patches tending to coalesce, only 40% of leaf affected	12
3	60% of leaf with feeding marks as chlorotic patches	107
4	Dense feeding marks over entire leaf, but appearance still green	228
5	As 4 but leaf blanched and starting to shrivel	592

The economic threshold of damage on established cucumber plants, above which crop loss occurs, is 1.9, while an index of 2.5 will cause a 40% loss after 5 weeks (the time required for cucumber fruits to develop). On cucumbers, the maximum rate of increase in leaf damage is 1.0 in 12 days.

On tomatoes, a similar threshold (2.0), equivalent to about 30% of photosynthetic area, initiates crop loss, while the maximum rate of increase recorded is 2.7 in 16 days. It is interesting to note that the rate of damage increase is similar in both tomatoes and cucumbers although the mean growing temperatures are 16 and 21°C, respectively.

Recently a new form of damage caused by a different red spider mite, *Tetranychus cinnabarinus* (Boisduval), which is plum red in colour, has appeared in several parts of the UK and the Netherlands (Foster & Barker, 1978). The typical speckling symptoms do not appear but, instead, infested leaflets become prematurely chlorotic with small transparent lesions. The damage is caused by remarkably few mites and closely resembles magnesium deficiency. Bright yellow patches develop on the leaflets and necrotic patches from these gradually coalesce until the whole leaf withers and dies (Figure 15). Darker stripes may be seen leading from infested leaflets into the stipe, though the main stem is not affected. The characteristic feeding marks associated with red spider mites may or may not be present. If left unattended, the lower leaves wither and die rapidly and total death of plants has been observed. The damage seems to be caused by injected toxin which experiments at the GCRI confirmed were neither systemic nor viral.

Under natural conditions, *T. urticae* overwinters as diapausing mated

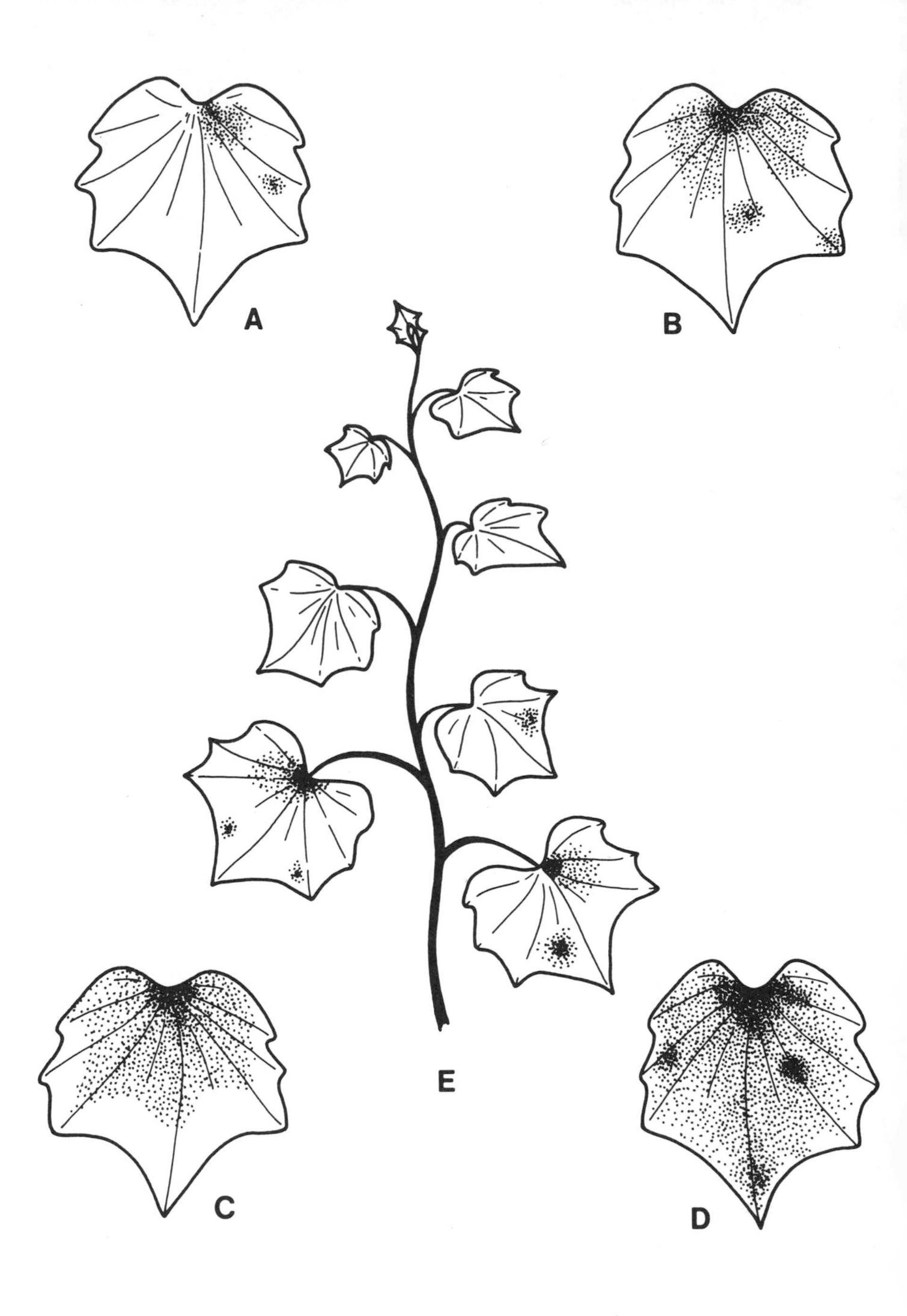

FIGURE 14 Indices for leaf damage by spider mites on cucumbers (*above*) and tomatoes (*right*). A = 1.0; B = 2.0; C = 3.0; D = 4.0; E = 5.0. On the cucumber plant, E illustrates a *mean* damage index of 0.4.

A
B
C
D
E

females, diapause being induced by shortening daylength, unfavourable food supply and low temperatures. Diapause is not normally terminated until the following spring when favourable conditions return. Both protonymphs and deutonymphs are sensitive, and respond, to shortening daylength.

The diapausing female spider mite is typically deep orange-red in colour – this colour being assumed within 3–5 days of maturation. Once this colour has developed, feeding ceases and the mites migrate from the plants to seek winter hibernation sites, usually within cracks and crevices in the greenhouse structure.

Daylength is important in controlling the onset of diapause and, in Southern England, the critical daylength is about 13½ hours, this period decreasing by 1 hour for each 3° fall in latitude. The intensity of illumination needed to induce a photoperiodic response in the developing nymphs, is about 3.5 lux. As with certain plants, breaks in the long dark cycle reduces the incidence of diapause, so long as the break is at least 2 hours and each portion of the broken dark period is less than 8 hours (Hussey, 1972). This response enables growers to set up more economic lighting regimes by reducing the current load.

Low temperatures, as may be experienced in the greenhouse when the heating is turned off, will also favour the onset of diapause, as will the availability of only a poor food source on senescing leaves. This occurs commonly, in September, on old main-crop cucumber plants, whereas younger second-crop plants growing at the same time carry few diapausing mites.

Once diapause has been induced, the females become positively geotactic and negatively phototrophic. They thus show relatively little lateral dispersal so that, each year, new infestations tend to arrive in the same part of the greenhouse.

Diapause is normally terminated only by a fixed period of chilling – an adaptation to ensure that mites do not become prematurely active on warm winter days. In the greenhouse, emergence usually begins when the heat is turned on for a new season's crop, although this is not always the case. Extensive research in the USSR (Hussey, 1972) has revealed that the pattern of emergence from diapause is governed by the combination of factors which induced hibernation in the first place. Hence, in some years, emergence has been delayed until April or even May, despite the fact that the greenhouse has been heated for several weeks. The scale of an infestation in the spring is, therefore, governed by the number of mites which enter diapause in the late summer.

Sex determination is arrhenotokous or haplodiploid (females develop

FIGURE 15 Serious defoliation on tomato caused by 'hypertoxic' strain of *Tetranychus cinnabarinus*.

from fertilized eggs and have a normal complement of chromosomes). The male on the other hand has only one set of chromosomes so that mutations will be immediately expressed. Thus, there is a rapid interaction between mutation and natural selection, enhancing the potential for rapid development of immunity to pesticides.

2.3b **THE RED SPIDER MITE PREDATOR (*PHYTOSEIULUS PERSIMILIS*)** This predator was originally known as *P. riegeli* Dosse and *P. tardi* Lombardini (Plate 5). It was accidentally brought from Chile to Germany on orchid roots and subsequently sent to many other parts of the world. Since the early 1960s, extensive research has been conducted both in Europe (Bravenboer & Dosse, 1962) and North America, and its ability to control *T. urticae* on a wide range of host plants has been clearly demonstrated.

The adult female is an orange-red, pear-shaped mite rather longer than its prey. The nymphs are oval and very pale pink in colour. Females lay eggs singly on the undersides of leaves among colonies of *T. urticae*. The eggs are large (2 × size of *T. urticae* eggs). When first laid they are translucent turning pink-orange. After about 3 days (at 20°C) a six-legged larva hatches and a day later it moults into an eight-legged protonymph which actively searches for food, eating 4–5 eggs before developing into a deutonymph. The deutonymph stages last about 2 days during which some 6 eggs or young mite stages may be eaten. Adults eat about 7 mites/day. The sex ratio (♀:♂) is approximately 4:1 and when egg-laying begins prey consumption doubles. Oviposition continues for about 3 weeks at a rate of 2–3 eggs/day (total about 54). Unmated females will not reproduce.

Under experimental conditions, at 20°C, the predator population increases 44 × in a mean generation time of 17 days. This is equivalent to a weekly increase of 4.6 × compared with that of 2.7 × by its host. Given these statistics, it is not surprising that *P. persimilis* is a most effective predator and has been considered by some to be too effective as it often eradicates its prey from a greenhouse.

At 20°C, a 300 × population increase occurs in 30 days while at 26°C this figure rises to about 200 000 × . Above 30°C, predators do not thrive though *T. urticae* continues to develop rapidly. In greenhouses, temperature is perhaps the most important factor governing the time required to achieve control of mite populations (see Figure 16). This becomes particularly important when cool outdoor conditions demand that heating the greenhouse temperature above ambient depends solely on pipe-heat. Spatial variations within the greenhouse caused by the pipe lay-out create differences in the speed and efficiency of predator control.

Various studies have shown that decreasing the relative humidity increases its searching, feeding and egg-laying capacity, although

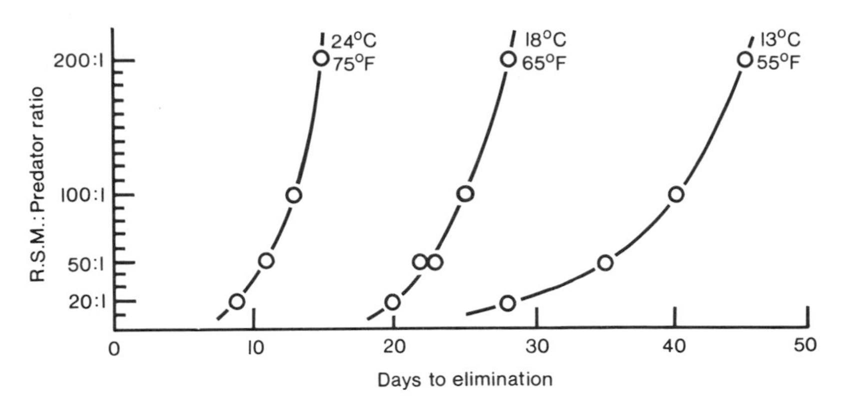

FIGURE 16 Time taken by predator to control red spider mite at different temperature and host-predator ratios.

predator development ceases below 60% RH when oviposition and longevity also decline sharply (Pralavorio & Rojas, 1980). This explains the downward migration of predators from the apical foliage of cucumbers in hot sunny weather and may justify the selection of more tolerant strains which have been obtained following extensive experiments in Leningrad. Here, Voroshilov (1979) claims to have increased the heat tolerance of certain strains by 8–10 × . In practice, migration and extensive searching for prey occurs only when the latter becomes scarce within the immediate vicinity.

Several species of phytoseiid mites have been the subject of genetic improvement projects and high levels of tolerance to parathion have been obtained with *P. persimilis.* Strains tolerant to carbaryl (× 10), diazinon and pyrazophos have been detected in Europe, while a project is underway in New Zealand to obtain pyrethroid resistance. Attempts to select for diazinon and pyrazophos resistance have been successful on a laboratory scale but less so in mass culture.

REFERENCES

Anon. (1978) The biological control of cucumber pests. *Glasshouse Crops Res. Inst. Growers' Bull.* No. 1: 19 pp.

Bravenboer, L. & Dosse, G. (1962) *Phytoseiulus riegeli* Dosse als Prädator einiger shadmilben aus der *Tetranychus urticae* Gruppe. *Entomologia Exp. Appl.* **5**: 291–304.

Foster, G.N. & Barker, E. (1978) A new biotype of red spider mite causing atypical damage on tomatoes. *Pl. Path.* 27: 47–48.

Hoy, Marjorie, A. (1982) Recent advances in knowledge of the Phytoseiidae. Division of Entomology, University of California. Publication No. 3284: 92 pp.

Hussey, N.W. (1972) Diapause in *Tetranychus urticae* and its implication in greenhouse culture. *Acarologia* **13**: 344–350.

Hussey, N.W. & Huffaker, L.B. (1976) Spider mites. pp. 179–228. In *Studies in Biological Control* (Delucchi, V.L., ed.) Cambridge University Press. 380 pp.

Hussey, N.W. & Parr, W.J. (1963) The effect of glasshouse red spider mite (*Tetranychus urticae* Koch) on the yield of cucumbers. *J. Hort. Sci.* **38**: 255–263.

Laing, T.E. (1968) Life history and life tables of *Phytoseiulus persimilis*. *Acarologia* **10**: 378–388.

Pralavorio, M. & Rojas, A. (1980) Influence de température et de l'humidité relative sur le développement et la reproduction de *Phytoseiulus persimilis*. *Bull. IOBC/WPRS Working Group on Integrated Control in Glasshouses, Vantaa* **3** (3): 157–162.

Scopes, N.E.A. & Biggerstaff, S.M. (1973) Progress towards integrated pest control on year-round chrysanthemums. pp. 227–234. In *Proceedings of the 7th British Insecticide and Fungicide Conference* 1973.

Scopes, N.E.A. & Ledieu, M.S. (1979) Integrated control of tomato pests. *Glasshouse Crops Res. Inst. Growers' Bull.* No. 3: 28 pp.

Voroshilov, N.V. (1979) Heat-resistant lines of the mite *Phytoseiulus persimilis* A-H. *Genetika* **15**: 70–76.

2.4 THRIPS AND THEIR NATURAL ENEMIES
N.W. Hussey

Thrips are small, slender insects commonly called 'thunderflies' by country folk in the UK. Many are of tropical origin but have become widely established on ornamentals in greenhouses. They may cause serious damage to cucumbers, carnations, roses and other flowering plants.

2.4a **BIOLOGY** The life cycle comprises the egg, two larval, one prepupal and one pupal instar. The adults have two pairs of narrow wings fringed with long, fine hairs which, when at rest, are laid parallel along the back.

It is convenient to consider the commonest species, *Thrips tabaci*, as typical of the several other pest species.

T. tabaci is about 1 mm long and greyish yellow-brown in colour. The antennae are yellow-brown and seven-segmented. The legs are yellow, shaded with brown, the wings yellow brown. The younger stages are yellow-green. Males are very rare, the females reproducing parthenogenetically.

Under summer conditions, the females each lay about 60 whitish, uniform eggs singly within a slit cut by the ovipositor in leaf or flower tissues. One end protrudes from the slit to facilitate emergence of the newly hatched larva. The first stage is only 0.4 mm long and has bright red eyes on an abnormally large head. Larval development is completed in 10–14 days when the larva drop to the ground to form prepupae which, in less than 2 days, turn into a pupa. The antennae of the prepupae are, unlike the pupae, not turned back over the head. The adults develop within the pupae in 4–7 days. Normally, the pupae are found on the soil surface or in natural cavities up to 15 mm below the surface.

Thrips are gregarious and large numbers are found together on a single leaf or flower (Plate 6). They feed on sap after piercing tissues with their mouthparts. The tissues around the feeding punctures become desiccated, giving the leaves or petals a flecked appearance; indeed, on the rapidly expanding leaves of cucumbers, these flecks enlarge to become 'windows'. Thrips usually lie alongside prominent veins so that a concentration of damage occurs where the principal veins radiate from the petiole (Figure 17).

Thrips are very susceptible to pesticides so that they rarely become serious pests where fully chemical pest and disease programmes are used. However, where biological control of major pests is practised, thrips create serious problems as they occupy niches on the plant (more than 50% of the population on the lower surfaces with a tendency for

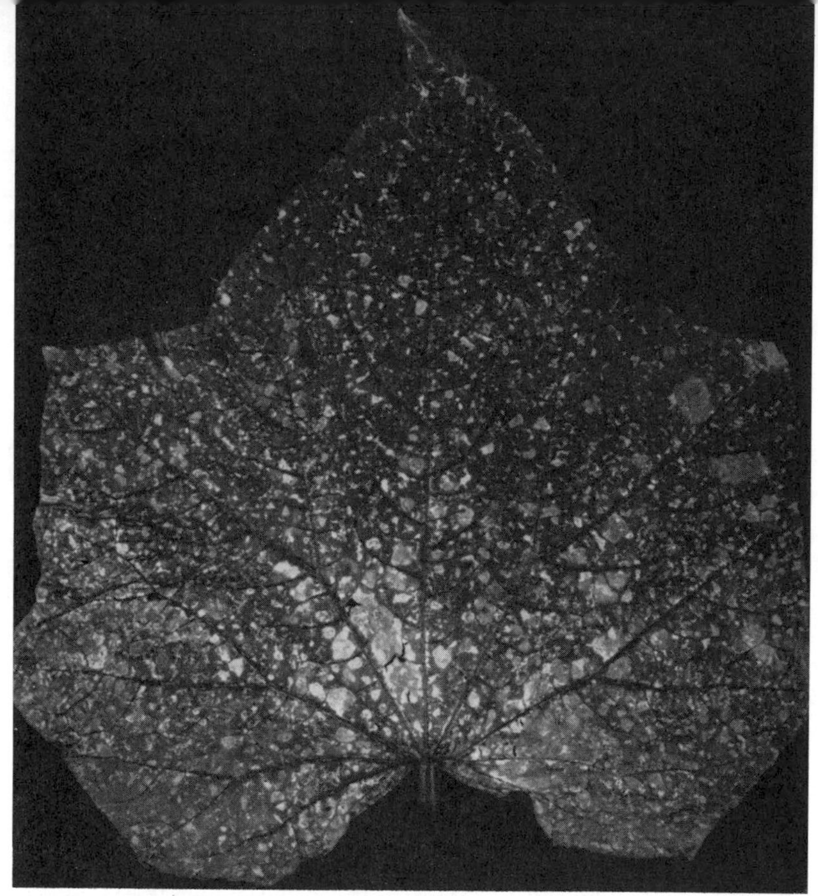

FIGURE 17 Leaf damage on cucumber caused by *Thrips tabaci*.

greater numbers to occur on the younger upper leaves) occupied by mites and whiteflies. It has, therefore, not proved possible to kill thrips on their host plants without affecting mite predators and whitefly parasites. Attention was therefore directed to the prepupae, which are found immediately below the plants, since larvae fall to the ground rather than walking down the main stem of the host.

In recent years, in the face of escalating fuel costs, greenhouse cultural techniques have changed dramatically. Soil sterilization by steam or methyl bromide is no longer practised and the roots are now protected from disease by the use of rock-wool or peat-bolsters. Thrips are, therefore, able to overwinter freely and now attack young plants in mid-winter whereas attacks formerly occurred only in early summer when the pest immigrated into greenhouses from outdoors.

2.4b **CONTROL** When cucumbers were cultivated in manure beds and hose-watered, the soil and pathways were kept very wet – indeed, frequently waterlogged – and it was possible to predict where thrip damage could be expected. If the paths were dry, symptoms were almost universally present but where they were wet no damage occurred. This effect was partially caused by drowning but was also due to the encouragement of fungal epizootics which are discussed later in this section.

Conversely, drip-watering associated with the new above-soil growing techniques led to dry floors on which thrips readily survived with a consequent high rate of population increase on the plants above. This development coincided with a trend to earlier planting so that the first pesticides (γ-HCH and diazinon), which had been used to successfully control pupating thrips, caused the death of parasites and predators of other pests when toxic vapours reached lethal concentrations in closed greenhouses. Pickford (1984) investigated a range of other non-volatile, persistent materials but only the novel 50/50 mixture of polybutene and water, to which the insecticide deltamethrin had been admixed, achieved control – reducing populations on the leaves from $40/cm^2$ to less than $2/cm^2$ within 4 weeks of treatment. Subsequent large-scale trials demonstrated that this control lasted 10 weeks. Where damage had occurred at the time of treatment, 'control' was apparently slow as damaged leaves remained on the plant but, in reality, thrip populations dropped sharply within a few days. The product concerned is now marketed as Thripstick® and this selective control method could well be used on other crops when it would pay to lay polythene sheeting along the plant rows to facilitate the treatment (Plate 7).

Before the development of the Thripstick® concept, considerable efforts were made to identify potentially useful biological control techniques for thrips. The Commonwealth Institute of Biological Control was commissioned to seek natural enemies in onion crops in Central and Southern Europe. The most interesting agent found was the fungus, *Entomophthora parvispora.* Outdoor epizootics were first found from early July but by September attacks were widespread.

Soon after the integument of the thrips is penetrated by the germ-tube of a conidial spore, the body becomes filled with rectangular hyphal bodies, which increase in number by fission and kill the host insect in 3–6 days. In nymphs, the entire body surface becomes covered with single, unbranched conidia, each bearing a single spore which is spherical but bears a knob-like projection. In the case of adults, this conidial growth is restricted to the intersegmental membranes, which are markedly stretched by the compact mass of hyphal bodies within. As autumn approaches, the hyphal bodies change into brown resting spores. Attacked individuals containing these spores turn black and are washed or blown from the foliage by winter storms and release the spores on the soil surface. Only these resting spores survive the winter.

Field studies suggested that, regardless of the suitability of the environment, epizootics are largely dependent on host density. This restriction is similar to that found in many *Entomophthora* sp. affecting aphids. There is, therefore, no prospect of using this species as a microbial insecticide but its presence in glasshouses is no doubt a contributory factor in depressing population increases in cucumber crops grown in damp soil.

Samson *et al.* (1979) found another species, *E. thripidum*, in Dutch cucumber-houses. The fungus was not found before mid-August and epizootics did not occur before mid-October. This species differs from *E. parvispora* in that the spores are broadly ellipsoidal with a broad, truncate base and pointed apex. Long sporophores appear along the intersegmental membranes from which mature spores are forcibly ejected. Before the fungus sporulates, infected thrips move to an elevated part of the leaf to facilitate spore release. The fungus completes its life cycle in the insect host within 4 days. Where this fungus occurs, it can almost eliminate a thrip population within 2 weeks. In dull weather, sporulation occurs continuously but it ceases in bright sunshine.

This fungus has not been isolated in pure culture and so any development of microbial control will probably depend on *Verticillium lecanii*, which also attacks thrips (Binns *et al.*, 1982). In glasshouse experiments on cucumbers, *Verticillium* killed more than 80% of the thrips within 6 days (Gillespie *et al.*, 1983). In view of the commercial development of this fungus for control of aphids and whiteflies, the potential for another specific product is evident.

However, it is the development of predators for thrip control which has excited most interest. As the result of a planning decision within the Working Party, it was decided to concentrate biological control work on thrips at Naaldwijk while GCRI investigated the chemical approach. Studies by Ramakers (1980, 1983) have demonstrated the potential of the predatory mites, *Amblyseius mackenziei* and *A. cucumeris*, for thrip control, especially in situations where their reproductive rate is somewhat reduced by damp growing conditions.

These predators are pear-shaped, pale whitish brown and active mites. They are noticeably smaller and flattened compared with *Phytoseiulus*. They lay smaller eggs which are white whereas the eggs of spider mite predators are tinted with brown. Another difference is that *Amblyseius* eggs are often attached to plant hairs.

In experiments where 28 predators were released on 2 out of 17 cucumber plants, *Amblyseius* spread to all the plants within 5 weeks and a sharp decline in the thrips population was observed after 8–10 weeks.

A major advantage of this predator is that, unlike *Phytoseiulus*, it can survive the absence of its prey by taking other food, such as spider mite nymphs. Hence, although the numbers of predators declines after control has been achieved, the proportion of leaves on which *Amblyseius* can be found remains high for some weeks. Although both *Amblyseius* spp. and *Phytoseiulus* prey upon each other to some extent, they are able to co-exist.

Undoubtedly, the main interest in these predators lies in the ease with which they can be mass-produced (Ramakers & van Lieburg, 1982). Both species are reared on a flour mite, *Acarus farris*, which itself feeds on wheat bran within a stainless steel drum which is slowly rotated and within which the humidity can be accurately controlled. Up to 100 000 predators can be produced per litre of rearing volume on food costing less than £0.10p.

REFERENCES

Binns, E.S., Hall, R.A. & Pickford, R.J.J. (1982) *Thrips tabaci* distribution and behaviour on glasshouse cucumbers in relation to chemical and integrated control. *Entomologists Mon. Mag.* **188**: 55–68.

Carl, K.P. (1975) An *Entomophthora* sp. pathogenic to *Thrips* spp. and its potential as a biological control agent in glasshouses. *Entomophaga* **20**: 381–388.

Gillespie, A.T., Hall, R.A. & Burges, H.D. (1983) Control of onion thrips, *Thrips tabaci*, with entomogenous fungi. *Rep. Glasshouse Crops Res. Inst.* **1981**: 119–120.

MacGill, E.I. (1927) The biology of Thysanoptera with reference to the cotton plant. 2. The relation between temperature and life-cycle in a saturated atmosphere. *Ann. Appl. Biol.* **14**: 501–512.

MacLeod, D.M., Tyrrell, D. & Carl, K.P. (1976) *Entomophthora parvispora* a pathogen of *Thrips tabaci*. *Entomophaga* **21**: 307–312.

Pickford, R.J.J. (1984) Evaluation of soil treatment for control of *Thrips tabaci* on cucumbers. In *Tests of Agrochemicals & Cultivars Ann. Appl. Biol.* **5**: 18–19.

Ramakers, P.M.J. (1978) Possibilities for biological control of *Thrips tabaci* in glasshouses. *Meded. Rijksfac. Landb. Gent* **43**: 463–469.

Ramakers, P.M.J. (1980) Biological control of *Thrips tabaci* with *Amblyseius* spp. *Bull. IOBC/WPRS Working Group on Integrated Control in Glasshouses, Vantaa* **3** (3): 203–207.

Ramakers, P.M.J. (1983) Mass production and introduction of *Amblyseius mackenziei* and *A. cucumeris*. *Bull. IOBC/WPRS Working Group on Integrated Control in Glasshouses, Darmstadt* **6** (3): 203–206.

Ramakers, P.M.J., van Lieburg, M.J. (1982) Start of commercial production and introduction of *Amblyseius mackenziei* for the control of *Thrips tabaci* in glasshouses. *Meded. Rijksfac. Landb. Gent* **47**: 541–545.

Samson, R.A., Ramakers, P.M.J. & Oswald, T. (1979) *Entomophthora thripidum*, a new fungal pathogen of *Thrips tabaci*. *Can. J. Bot.* **57**: 1317–1323.

2.5 BIOLOGY OF GLASSHOUSE LEAF-HOPPER AND ITS PARASITE

M.J.W. Copland & Woeriani Soeprapto

2.5a ***ZYGINA PALLIDIFRONS*** While several native leaf-hoppers may occasionally occur in greenhouses in Europe, the most important is the greenhouse leaf-hopper, *Zygina palidifrons* Edw. This species was first recorded as a greenhouse pest in the UK in 1918 and increased rapidly during the period 1920 to 1940 (Fox-Wilson, 1938). These leaf-hoppers are relatively easily controlled using modern pesticides but, with the increasing use of biological control methods for other pests, they are once more becoming important on many holdings.

Z. pallidifrons is a small (3–4 mm long) pale-coloured jassid which is long-lived and extremely active, capable of developing serious infestations both under summer greenhouse conditions and within cages in controlled environment conditions.

It has a wide host range including tomato, cucumber, geranium, *Nicotiana*, french bean, fuchsia, chrysanthemum, cotton and hop among the plants attacked. Several overlapping generations occur through the summer and the leaf-hoppers can overwinter on weeds such as chickweed. Leaf-hoppers feed only on the leaves; stems or buds are never touched. The characteristic damage appears as a mottled area on the upper leaf surface (Figure 18) caused by individual leaf-hoppers feeding from the underside. In extreme cases, the mottled areas become confluent, rendering the leaf white, bleached and shriveled. All the young developmental stages resemble the adult and cause a similar type of damage. Smith (1926) observed that the feeding punctures continued to enlarge even after the removal of the insect because damaged cells around the feeding puncture collapse and become full of air. Our observations showed that the area of damage varied depending on host plant. Whether this reflected host preferences or cell size within the leaf tissue was not determined. At 24°C, most feeding occurred on cucumber (17 mm^2/24 hours) and least on *Nicotiana* (4 mm^2/24 hours). Temperature also affects the rate of feeding. The area of damage caused by feeding punctures on tomato at 26°C (14 mm^2/24 hours) was approximately twice that of the area affected at 18°C (5 mm^2/24 hours). Very little feeding occurs at temperatures below 14 or above 30°C. Provided with water, but no food, leaf-hoppers kept at 24°C survived only 3 days.

FIGURE 18 Spotting on tomato leaf caused by feeding of glasshouse leaf-hopper, *Zygina pallidifrons*.

FIGURE 19 Cast skins of leaf-hopper or 'ghost fly'.

The feeding punctures disfigure ornamental subjects and weaken or stunt seedlings. In addition, the leaf-hoppers produce honeydew which supports the growth of sooty moulds and hence may create a problem on food crops such as tomatoes.

Development of the leaf-hopper has been reported as lasting from 25 days in summer to 85 days in winter (MacGill, 1932). Our own studies have been made at four temperatures and show that, as with other insects, development is greatly influenced by temperature. There are five nymphal instars. The moulted skins characteristically remain attached to the plant by the stylets and, as they are white, have been referred to as 'ghost flies' (Figure 19). At 18°C, egg incubation takes approximately 17 days while the life cycle to adult emergence is complete in 42 days. At 30°C, egg incubation takes 8 days and adults emerge in 23 days. However, this temperature is very nearly at the upper lethal level for development so that few individuals mature successfully. Adults live for a long period, MacGill (1932) recording a life of up to 4 months. At temperatures near to 35°C, however, they die in a few hours.

Adults readily jump and fly. When a severely infested plant is disturbed, they fly up and hover temporarily in a similar manner to whitefly. The nymphal stages, however, are relatively slow moving and appear to spend most of their development on the leaf on which they hatched. The adults are able to locate and colonize other host plants with remarkable ease, dispersing readily around the greenhouse. Below 14°C, very little activity takes place and leaf-hoppers become sluggish.

The sexes are easy to distinguish, the females having an obvious ovipositor on the ventral surface. Males and females occur in equal numbers, although the females appear to be rather longer lived and hence dominate in severe infestations. Females are not sexually mature until a week or more old, although mating and courtship have not been observed in detail. In experiments to determine the age at which oviposition begins, newly emerged females were confined in clip cages on leaves with males. Eggs are nearly always laid singly, the leaf-hoppers preferring to oviposit in secondary rather than major leaf veins. In many cases, no eggs were laid, suggesting a reluctance to oviposit or perhaps mate under these conditions. As with other Hemiptera, egg production appears to be inversely related to temperature. Individual adults maintained at low temperatures contained more eggs in the ovaries than those kept at high temperatures. More eggs were laid at 18°C than at 30°C. We found it particularly difficult to find eggs laid within the leaf veins of cucumber hosts and usually waited for the egg to hatch. In general, the lower fecundity, the long maturation period, an unwillingness to mate, oviposition site discrimination and the difficulty in detecting newly laid eggs make experimental procedures fairly challenging.

2.5b **THE PARASITE *ANAGRUS ATOMUS* HALIDAY** The principal parasite of greenhouse leaf-hopper is the mymarid wasp, *Anagrus atomus* (see MacGill, 1934). As with other mymarids, this wasp is a parasite of the egg stage of their leaf-hopper host. Mymarids are minute, short-lived wasps with a complement of fully mature eggs which are laid as quickly as possible. While the parasites seem readily able to find and attack leaf-hopper eggs, the percentage of parasitized hosts remains small. At 24°C, development take about 16 days. Towards the end of parasite incubation, parasitized eggs may be clearly seen as they develop a reddish hue. Parasites of such an early developmental stage are unable to exercise effective control of the leaf-hopper. We believe that an attempt to achieve biological control using this mymarid alone is unlikely to be successful. While this wasp can be bred relatively easily in culture, there are several handling problems; eggs desiccate within detached leaves maintained for parasite emergence and the adult wasp is too short-lived for distribution to nurseries.

REFERENCES

Fox-Wilson, G. (1938) The glasshouse leafhopper, *Erythroneura pallidifrons* Edw. *J. R. Hort. Soc.* **43**: 481–484.

MacGill, E.I. (1932) The biology of *Erythroneura pallidifrons* Edw. *Bull. Ent. Res.* **23**: 33–43.

MacGill, E.I. (1934) The biology of *Anagrus atomus* Hall., an egg parasite of the leafhopper *Erythroneura pallidifrons* Edw. *Parasitology* **26**: 57–63.

Smith, K.M. (1926) A comparative study of feeding methods of certain Hemiptera and of the resulting effects upon the plant tissue, with special reference to the potato plant. *Ann. Appl. Biol.* **13**: 109–139.

2.6 LEAF-MINERS AND THEIR PARASITES

L.R. Wardlow

2.6a **THE THREE MOST TROUBLESOME LEAF-MINER SPECIES** (*Liriomyza bryoniae*) TOMATO LEAF-MINER attacks both tomatoes and cucumber and will survive on some outdoor weeds, e.g. sowthistle. The pest is important where greenhouse soil is not sterilized as leaf-miner pupae can then survive in the ground from one crop to the next. Crops grown by the nutrient-film method (NFT) or in peat bags are particularly susceptible to this pest, especially where the floor is covered with plastic sheeting.

(*Chromatomyia syngenesiae*) CHRYSANTHEMUM LEAF-MINER has a wide host range including many weeds; it is a severe pest of chrysanthemums with some cultivars being particularly prone to attack. This species pupates inside the leaf where it is difficult to control with insecticides.

(*Liriomyza trifolii*) AMERICAN SERPENTINE LEAF-MINER also has a wide host range and attacks most major protected crops. Like tomato leaf-miner, this species also pupates on the ground and is likely to thrive under similar conditions. The pest was imported into Europe from the USA comparatively recently via various African and Mediterranean nurseries, where it is now a serious problem due to its resistance to insecticides. In the UK, it is a notifiable pest under Statutory Plant Health Regulations and outbreaks have to be eradicated as quickly as possible. Insecticides currently recommended to control this pest would seriously interfere with biological control programmes for other pests.

2.6b **LEAF-MINER BIOLOGY** Tomato and American serpentine leaf-miner adults are difficult to distinguish with the unaided eye. Both are small black flies (2.5 mm long) with yellow spots on the thorax between their wings (Plate 8). It is most important for nurserymen to have a correct identification made of the species on their crops. Chrysanthemum leaf-miner is larger (3 mm long) and a dark grey colour (Plate 9).

Adult leaf-miners feed on the leaf sap produced when the female fly inserts her ovipositor into the upper leaf surface to seek a suitable site for inserting an egg within the tissue. These oviposition marks soon show up as pale white spots on the upper surfaces of the leaves. Only a proportion of the punctures are eventually chosen as egg-laying sites. Each female lays an average of 60 eggs during a 2–3 week life. After about 1 week the eggs hatch into small translucent larvae which tunnel within the leaf tissue. Each larva forms a pale white tunnel or mine. The larvae feed for up to 10 days, moulting 3 times, and as they become larger the mine also

FIGURE 20 Mines of *Chromatomyia syngenesiae* (*left*) and *Liriomyza trifolii* (*right*) in chrysanthemum leaves.

becomes larger, longer and broader. The mines usually meander through the leaf but the American leaf-miner may also cause conspicuous sharply curled or blotched mines (Figure 20). Mature larvae emerge from the mine to pupate (falling to the ground in the case of both *Liriomyza* species) for about 9 days after which the adult flies emerge to repeat the cycle. Leaf-miners are capable of breeding throughout the year in heated greenhouses and are deceptive in their capacity to reproduce to outbreak proportions.

2.6c **LEAF-MINER PARASITES** There are two distinct types of parasite with many species attributable to each. Three species are available commercially but several other species may occur naturally where commercial biological control of leaf-miners is practised.

Endoparasites lay their eggs inside the leaf-miner larvae. After hatching, the parasite larvae feed at a rate which ensures that the leaf-miner is not killed until it has pupated. Only 1 parasite matures within each leaf-miner larva, although more than 1 parasite egg may be laid.

Ectoparasites lay their eggs alongside the leaf-miner larva within its mine (Plate 10) but the leaf-miner is first paralysed by the adult so that it cannot move away from the parasite egg; the ectoparasite larva then feeds on the leaf-miner externally. Up to 6 parasites have been found developing within 1 mine, although the adults subsequently vary in size. Once a leaf-miner is attacked by an ectoparasite, the mine develops no further and, if it is a small young mine, it may not be detectable as the leaf grows.

2.6d **COMMERCIALLY AVAILABLE PARASITES** *Dacnusa sibirica* (Braconidae) is a small black endoparasite (2–3 mm long) with long flexible antennae (same length as its body). Parasites are supplied to nurserymen either as pupae in small cardboard release-boxes or as adults in plastic tubes. *Opius pallipes* and other species of *Dacnusa* may also be found amongst commercial supplies but they operate in a similar manner. The female parasite uses her antennae to locate a leaf-miner larva within a leaf, whereupon she inserts a transparent oval egg with her ovipositor. Each female lives for about 2 weeks, during which time she may lay up to 90 eggs. Eggs hatch within 4 days, the larvae taking about 16 days to mature within the leaf-miner pupae. With the aid of a transmitted light microscope, it is possible to dissect out leaf-miner larvae from the leaves before pupation to check parasitism.

A method for mass-rearing *Dacnusa* spp. is described by Hendrikse (1980) and another for *Diglyphus* by Hendrikson (1975).

Diglyphus isaea (Eulophidae) is an ectoparasite and is common on outdoor weeds from June onwards in the UK. The adult is small (1–2 mm long) and black with a metallic green sheen; its antennae are much shorter than those of *Dacnusa sibirica*. Little is known about the fecundity of *D. isaea* but a female probably lays about 60 eggs during her life span (Merrett, 1978). When the parasite has paralyzed the leaf-miner, she inserts an egg through the leaf, placing it adjacent to the host. The parasite larva hatches within 2 days, passing through three growth stages during the next 6 days (Ibrahim & Madge, 1979). Before turning into a pale turquoise pupa, the larva builds 6 to 8 columns with frass within the mine; these act like 'pit props' to prevent damage to the pupa. The pupa then turns dark brown before the adult emerges through a hole chewed in the leaf 6–9 days later. The frass columns remain in the mine and are a useful indication of the degree of parasitism. *Diglyphus* spp. are able to parasitize leaf-miners already parasitized by *Dacnusa* spp. and hence they become the dominant parasites when the hot conditions of mid-summer favour their activity. Adult *Diglyphus* also feed on leaf miner larvae to obtain protein – an essential ingredient of their diet to maintain egg production.

REFERENCES

Hendrikse, A. (1980) A method for mass-rearing two braconid parasites (*Dacnusa sibirica* and *Opius pallipes*) of the tomato leaf miner (*Liriomyza bryoniae*). *Meded. Rijksfac. Landb. Gent* **45** (3): 563–571.

Hendrickson, J.M. (1975) Mass rearing of *Diglyphus isaea* Walker (Hymenoptera: Eulophidae). *J. N.Y. Ent. Soc.* **83**: 243–244.

Ibrahim, A.G. & Madge, D.S. (1979) Parasitization of the chrysanthemum leaf miner *Phytomyza syngenesiae* by *Diglyphus isaea. Entomologists' Mon. Mag.* **114**: 71–82.

Merrett, P.J. (1978) The chrysanthemum leaf miner *Chromatomyia syngenesiae* and its ectoparasite *Diglyphus isaea.* Ph.D. Thesis, University of Reading, UK.

2.7 BIOLOGY OF APHIDS AND THEIR PARASITES IN GREENHOUSES

J.M. Rabasse & I.J. Wyatt

Aphids rank with spider mites and whiteflies as the most serious pests of greenhouse crops. There are many more species than of the other pest groups but, until recently, they have proved more amenable to chemical control. However, resistant strains are appearing and, with increasing use of biological control against the other pests, it is becoming more necessary to find aphid control measures which are compatible with biological control.

2.7a **BIOLOGY OF GREENHOUSE SPECIES** **Morphology** Aphids (Homoptera, Aphididae) are small (2 mm) soft-bodied insects which live on plants in dense colonies. The body is generally pear-shaped and lacks obvious segmentation or division into head, thorax and abdomen. The legs and antennae are slender and the hind end of the abdomen bears a pair of tubular wax glands, the cornicles. Two forms of adult occur: the apterae, which are wingless, and the alatae, which are winged. The latter appear particularly in crowded conditions.

Feeding Aphids feed on plant sap by inserting the slender stylets of their mouthparts through the plant tissue and into the sieve tubes. Since the phloem sap is low in amino acids but rich in sugars, the aphids must take in large quantities of sap to meet their amino acid requirements, and pass out the remaining sugary liquid, or honeydew, droplets of which are ejected for considerable distances around the aphid. The honeydew not only clogs the stomata of the leaves, but encourages the growth of sooty moulds which prevent light from reaching the photosynthetic tissues. The aphids weaken the plant by draining its resources and may cause severe distortion of growth and, additionally, are a common means of transmitting plant viruses from infected to healthy plants.

Host range Aphids are highly adapted as plant parasites, showing their widest diversity and specialization in temperate regions. Most groups of plants have one or more aphids specific to them. On the other hand, some species infest a wide range of hosts. Many aphids alternate between summer host plants and a woody winter host on which they reproduce sexually and lay overwintering eggs. However, under glass, reproduction is continually asexual even through the winter or, as with the rose aphid (*Macrosiphum rosae* L.) the rose is strictly its winter host. No aphid species is confined entirely to the greenhouse, although a strain of *Myzus persicae* Sulz. has become highly adapted to chrysanthemums and to many insecticides applied to them, and is probably confined to greenhouse culture (Wyatt, 1966).

Aphids can thus be considered under two categories: the polyphagous species, attacking a wide range of hosts, and the oligophagous species, confined to a single host. The most important polyphagous species are *M. persicae* (Figure 21), which infests mainly Solanaceae and chrysanthemums, *Aphis gossypii* Glover, on cucumbers and chrysanthemums, and *Macrosiphum euphorbiae* Thom. and *Aulacorthum solani* Kalt., which are chiefly pests of Solanaceae, but, like the previous two species, can affect many other plants. *Aulacorthum circumflexum* Bckt. attacks a wide range of ornamentals, particularly under glass. Oligophagous species include *Nasonovia ribisnigri* Mosl., *Hyperomyzus lactucae* L. and *Acyrthosiphon lactucae* Pass. on lettuce, *Dysaphis tulipae* B.d.F. and *Myzus ascalonicus* Donc. on bulbs and *Macrosiphoniella sanborni* Gill. (Plate 11) and *Brachycaudus helichrysi* Kalt. (Plate 12) on chrysanthemums.

Reproduction The reproduction of aphids is highly adapted to exploit a new temporary habitat by rapid population increase; they are what are known as 'r-strategists'. They achieve this end in several ways. On greenhouse crops, as on other summer hosts, all individuals are females and therefore all contribute to population growth. The young are born fully formed and able to feed immediately. They grow rapidly, moulting 4 times before they mature, often within a week or less. Because fertilization is not required, ova can start developing within an aphid as soon as, or even before, it is born. By the time a female matures, several young are fully developed in her oviducts and ready to be born. Young are then produced at a rate of 3 or even 6 a day for several weeks.

The body structure of aphids is considerably simplified to perform only the functions of feeding and reproduction, while retaining the ability to walk. Even wings and flight muscles are dispensed with, except when

FIGURE 21 Adult and nymphs of *Myzus persicae*.

these are needed to escape to a new food source. Thus all nutrition is directed to the needs of reproduction.

2.7b **DEVELOPMENT OF APHID POPULATIONS** **Dispersion and establishment** Because aphids increase so rapidly on short-lived hosts, they must have mechanisms to prevent overcrowding and to disperse to new hosts. They therefore react to their own population density by both walking and flying off and by restricting their own reproductive rate (Wyatt, 1965). The initial invasion of a greenhouse in the spring is therefore often by outdoor alate migrants entering the vents. These may be aphids leaving their winter hosts, or flying between summer hosts, but flight cannot occur until outdoor temperatures are adequate. Thus, in 1982, the first *M. persicae* were trapped, from the South to the North of France, at Pau by 11 April, Orleans by 9 May, Colmar by 6 June and at Arras by 4 July (data from ACTAPHID suction trap network). Immigration may also be influenced by the size of vents and their time of opening.

Aphids may also be distributed between and within nurseries on propagating material. This is particularly the case with chrysanthemums and other crops where cuttings are supplied by a few specialist producers. Several crops such as lettuce, chrysanthemums, carnations and roses, are grown in greenhouses throughout the year. Aphids may

FIGURE 22 Evolution in time of the spatial dispersion (*top*) and of the frequency distribution (*bottom*) of aphid populations per plant in a greenhouse.

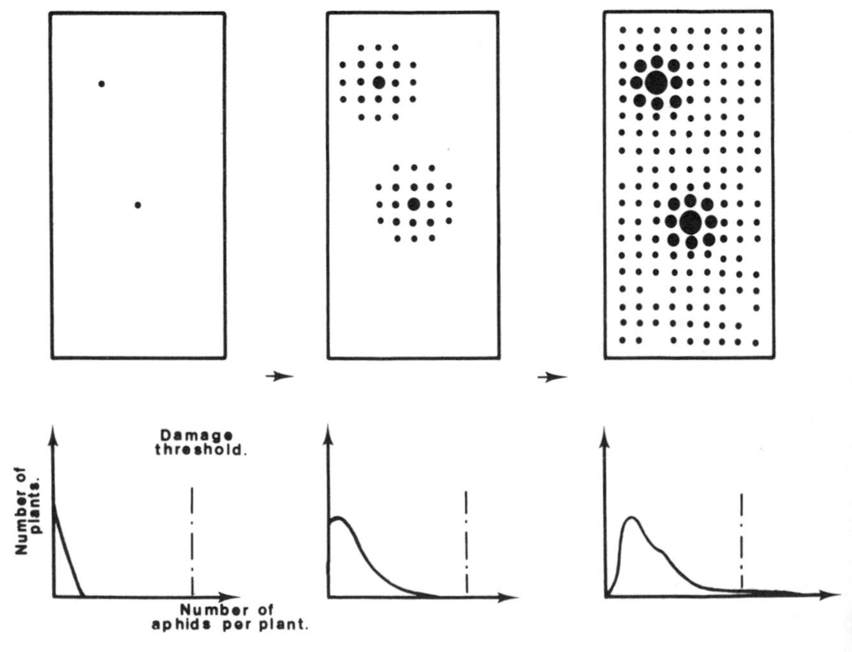

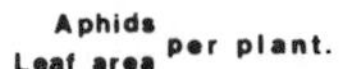

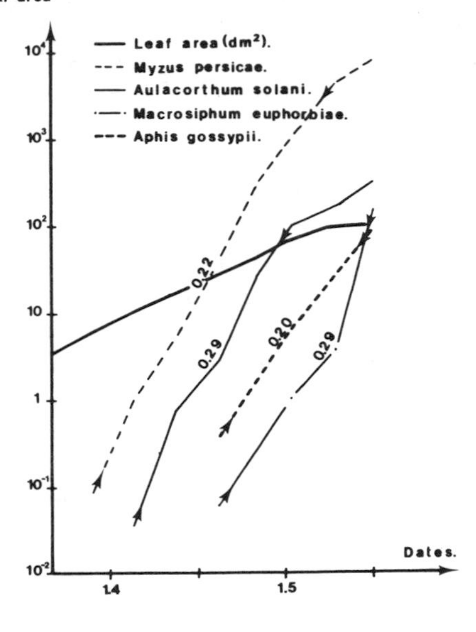

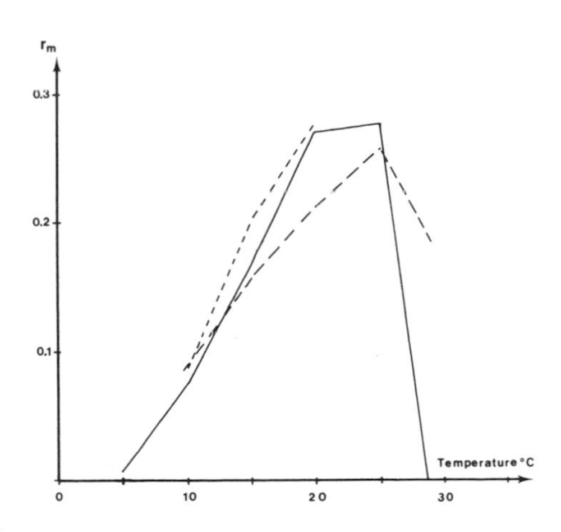

FIGURE 23 (*Top*) An example of population trends of several aphid species in a greenhouse (data from Rabasse, 1980). When the growth is more or less exponential, i.e. between arrows, the increase rate is mentioned.

FIGURE 24 (*Above*) Intrinsic rate of increase (r_m in females/female/day) for *Myzus persicae* Sulz. at different temperatures. (Data from ——— De Loach (1974); — — — El Din (1976); - - - Rabasse & Shalaby (1979).)

readily move from one crop to the next. Continuity may also result if weeds or old crops are left through the winter (Della Giustina, 1972a).

Distribution within a crop The initial infestation of a crop is usually at isolated foci. Rapid reproduction at these points produces dense colonies (Figure 22a) which, if recognized in time, can be controlled by localized treatments. Aphids soon begin to wander to neighbouring plants, however, particularly in species such as *Myzus persicae* where the adults are not inclined to settle (Figure 22b). As colonies become more dense, alatae are produced and disseminate the infestation throughout the crop (Figure 22c).

The distribution of aphids on individual plants depends on the aphid species and its density and on the species, cultivar and age of the plant (Figure 23). Thus, on Solanaceae, *M. persicae* prefers the lower leaves and *M. euphorbiae* the upper leaves. On chrysanthemums, *M. persicae* prefers the upper leaves, but moves to the middle leaves of susceptible cultivars as the population density increases, and then migrates to the flowers when they appear (Wyatt, 1965). In contrast, *M. sanborni* feeds on the stems of chrysanthemums while *Brachycaudus helichrysi* feeds in the growing point.

Population growth rates Within a greenhouse, environmental factors are constant and the natural enemies of aphids are usually absent. An aphid population is therefore able to grow exponentially for a considerable period, i.e. the numbers increase by a fixed proportion (q) each day. When the $\log_e$ of the number is plotted against time an approximately straight line results. Figure 22 illustrates such lines for four aphid species on aubergine at about 20°C (Della Giustina, 1972b). The slopes of these lines (calculated between the arrows) are known as the r_m values (0.22, 0.29, 020 and 0.29) and the antilog$_e$ gives the daily increase, q (1.25, 1.34, 1.22 and 1.34). A more meaningful expression of increase rate is the weekly increase, q^7, which gives values of 5, 8, 4 and 8 times a week respectively.

The rate of increase is determined by two factors of the aphid life history; the development time (d) from birth to first reproduction, and the numbers of female young (M) produced in an equivalent time (Wyatt & White, 1977). With aphids, all young are normally female. Thus increase rates can be calculated either from the slope of population graphs, as above, or from d and M, since the weekly increase rate

$$q^7 = M^{\frac{5.2}{d}}.$$

For example if an aphid first reproduces at 7 days old and then produces 20 young in the next 7 days, the weekly increase rate will be $20^{\frac{5.2}{7}}$ or 14.8 times.

The rate of increase is greatly influenced by factors such as temperature and host plant. For example, *A. gossypii* increases at only 4 × a week on aubergines, yet it will increase at 12 × on cucumber under glass and as much as 23 × under ideal laboratory conditions (Wyatt & Brown, 1977). The intrinsic increase rate (r_m) is proportional to temperature between two limits: a lower developmental threshold, usually near 0°C, and an upper lethal limit at about 25°C (Figure 24). This temperature is often exceeded under glass in southern countries, but aphids are able to survive if night temperatures are low enough (El Din, 1976; Rabasse, 1976).

2.7c **BIOLOGY OF PARASITES** All aphid parasites are Hymenoptera, or wasps in the broad sense, and belong to two families: the Aphidiidae, which are most important and are all aphid parasites, and the Aphelinidae which also parasitize other insects, such as scales and whiteflies.

The Aphidiidae include many important genera: *Aphidius*, *Praon*, *Ephedrus*, *Lysiphlebus*, *Monoctonus* and *Trioxys*. The adults are small (2 mm) slender wasps with black, brown, orange or yellow colouration. They live for only 1 or 2 weeks at 15–20°C when fed with honeydew or nectar. Their populations comprise rather more (60%) females than males. After mating, the female inserts her ovipositor into an aphid (Figure 25), usually by bending her abdomen forwards between her legs, and lays a minute (0.1 mm long) egg in its body cavity. Certain aphid instars are preferred by any one parasite species, although any instar is usually acceptable. Once within the aphid, the egg expands to several times its original size. After a few days, the larva hatches and begins feeding osmotically. The larva grows, passing through three instars,

FIGURE 25 Adult parasite *Asychis* sp. ovipositing in adult *Aphis gossypi*.

without interfering markedly with the development or behaviour of the aphid. By the 4th instar, the aphid has usually become adult and the parasite has consumed all the internal tissues of the aphid, completely filling its cuticle. The larva then cuts a slit in the underside of the aphid, attaches the cuticle to the leaf by silk and then spins a cocoon within the aphid, where it pupates. This is the 'mummy' stage and resembles a swollen, papery aphid, yellow, brown or sometimes black (*Ephedrus*) in colour (Plate 13). Parasites of the genus *Praon* spin a cocoon beneath the empty skin of the aphid. When mature, the adult parasite cuts a circular lid in the top of the mummy, leaving behind only a few pellets of meconium. At 21°C, the mummy of *Ephedrus cerasicola* Stary appears after 12 days and the adult 9.5 days later (Hofsvang & Hagvar, 1975). For *Aphidius matricariae* Hal. the equivalent times are 8.5 and 5 days (Scopes & Biggerstaff, 1977; Rabasse & Shalaby, 1980).

An adult female parasite may make several hundred oviposition attempts during its life, but only a small proportion of the eggs laid will develop successfully to adulthood. Under laboratory conditions up to about 100 adults will eventually be produced, of which perhaps 60 will be female (Vevai, 1942). Since development takes about 2 weeks the maximum population increase rate can be calculated as about 4.5 × a week, using the formula described for aphids. In practice, within a greenhouse, the increase rate may be considerably lower.

The Aphelinidae include one genus of importance: *Aphelinus*. The adults are small (1 mm) and thickset, the wings are short with a reduced venation, and the antennae are elbowed. The female inserts its ovipositor by backing up to the aphid. An egg may then be laid or the female may turn and feed from the puncture. Either process can lead to the death of the aphid. The mummy is black, retaining the original size and shape of the aphid, and the exit hole is ragged. Otherwise these parasites resemble the Aphidiidae in biological details such as development time and fecundity. The adults live somewhat longer.

REFERENCES

De Loach, C.J. (1974) Rate of increase of cabbage, green peach and turnip aphids at constant temperatures. *Ann. Ent. Soc. Am.* **67**: 332–340.

Della Giustina, W. (1972a) Integration de méthodes de lutte contre les ravageurs des plantes légumières cultivées en serre. Premiers résultats. *Anls Zool. Ecol. Anim.* **4**: 367–374.

Della Giustina, W. (1972b) Etude sur les fluctuations des populations d'insectes vivant dans les serres légumières de la region Parisienne. *Annls Zool. Ecol. Anim.* **4**: 5–33.

El Din, N.S. (1976) Effects of temperature on the aphid, *Myzus persicae* (Sulz.), with special reference to critically low and high temperature. *Z. Angew. Ent.* **80**: 7–14.

Hofsvang, T. & Hagvar, E.B. (1975) Developmental rate, longevity, fecundity and oviposition period of *Ephedrus cerasicola* Stary (Hym., *Aphidiidae*) parasitizing *Myzus persicae* Sulz. (Hom., *Aphididae*) on paprika. *Norw. J. Ent.* **22**: 15–22.
Hofsvang, T. & Hagvar, E.B. (1977) Cold storage tolerance and super-cooling points of mummies of *Ephedrus cerasicola* Stary and *Aphidius colemani* Viereck (Hym., *Aphidiidae*). *Norw. J. Ent.* **24**: 1–6.
Lyon, J.P. (1968) Remarques préliminaires sur les possibilités d'utilisation pratique d'Hyménoptères parasites pour la lutte contre les pucerons en serre. *Annls Epiphyt.* **19**: 113–118.
Rabasse, J.M. (1976) Note préliminaire sur l'utilisation des chocs thermiques en lutte intégrée contre *Myzus persicae* Sulz. en serre. *Bull. IOBC/WPRS Working Group on Integrated Control in Glasshouses, Antibes* **1976** (4): 99–103.
Rabasse, J.M. (1980a). Implantation d'*Aphidius matricariae* dans les populations de *Myzus persicae* en culture d'aubergines sous serre. *Bull. IOBC/WPRS Working Group on Integrated Control in Glasshouses, Vantaa* **3** (3): 175–185.
Rabasse, J.M. (1980b). Dynamique des populations d'aphides sur aubergine en serre. I – Considérations générales sur la colonisation et le développement des populations de quatre espèces dans le Sud de la France. *Bull. IOBC/WPRS Working Group on Integrated Control in Glasshouses, Vantaa* **3** (3): 187–198.
Rabasse, J.M. & Shalaby, F.F. (1979) Incidence du parasite *Aphidius matricariae* Hal. (Hym. *Aphidiidae*) sur le fécondité de son hôte *Myzus persicae* Sulz. (Hom., *Aphididae*) à différentes températures. *Annls Zool. Ecol. Anim.* **11**: 359–369.
Rabasse, J.M. & Shalaby, F.F. (1980) Laboratory studies on the development of *Myzus persicae* Sulz. (Hom. *Aphidiidae*) and its primary parasite, *Aphidius matricariae* Hal. (Hym. *Aphidiidae*) at constant temperatures. *Acta Oecol. Appl.* **1**: 21–28.
Scopes, N.E.A. & Biggerstaff, S.B. (1977) The use of a temperature integrator to predict the developmental period of the parasite *Aphidius matricariae*. *J. Appl. Ecol.* **14**: 799–802.
Vevai, E.J. (1942) On the bionomics of *Aphidius matricariae* Hal., a braconid parasite of *Myzus persicae* Sulz. *Parasitology* **34**: 141–151.
Wyatt, I.J. (1965) The distribution of *Myzus persicae* (Sulz.) on year-round chrysanthemums. I. Summer season. *Ann. Appl. Biol.* **65**: 439–459.
Wyatt, I.J. (1966) Insecticide resistance in aphids on chrysanthemums. pp. 52–55. In *Proceedings of the 3rd British Insecticide & Fungicide Conference 1965*. 519 pp.
Wyatt, I.J. & Brown, S.J. (1977) The influence of light intensity, daylength and temperature on increase rates of four glasshouse aphids. *J. Appl. Ecol.* **14**: 391–399.
Wyatt, I.J. & White, P.F. (1977) Simple estimation of intrinsic increase rates for aphids and tetranychid mites. *J. Appl. Ecol.* **14**: 757–766.

2.8 BIOLOGY OF THE MIDGE *APHIDOLETES* AND ITS POTENTIAL FOR BIOLOGICAL CONTROL

M. Markkula & K. Tiittanen

Little is known of those gall-midge species which are predatory on leaf-aphids. However, research on their biology has increased recently and it has become evident that the role of midges in regulating the abundance of aphids in nature is greater than has been previously assumed.

Three species are known to be predators of aphids: *Aphidoletes aphidimyza* (Rond.), *A. urticariae* (Kieffer) and *Monobremia subterranea* (Kieffer) (see Harris, 1973). Only *A. aphidimyza* has been investigated with respect to its suitability for the control of pest aphids in greenhouses.

A. aphidimyza is a holarctic species. It is known in the USA, Canada, Japan and in most European countries. For instance, in Finland, this species is very common and abundant, occurring far to the north, up to latitude 68°C (Markkula *et al.*, 1979).

2.8a **BIOLOGY OF *APHIDOLETES APHIDIMYZA*** The adult midge is about 2 mm long, slender with long legs. The sexes can easily be separated on the basis of the structure of the antennae. The male antennae are long, grey with long setae and bent backwards. The female antennae are shorter, thicker and darker in colour.

Monogenic reproduction is characteristic of the aphid midge, i.e. all progeny from a single female are either males or females (Sell, 1976). In one series of experiments, the ratio of males to females was 1:1.7.

The adults do not live long. Uygun (1971) observed that their life span averaged 1 week, the male living for a slightly shorter time. In the experimental glasshouses of the Agricultural Research Centre in Finland, they have survived for up to 2 weeks, but in the wild state their lives are much shorter, probably only a few days (Harris, 1973).

The adult midges feed on the honeydew secreted by aphids. They are active only at night and during dusk. In the daytime, they remain immobile in shaded parts of the plants.

The females oviposit, usually under the leaves, only on plants infested by aphids. Egg-laying is particularly prolific in aphid colonies where it is brought about by olfactory, chemical or tactile stimulation by the aphids, or their secretions, either alone or in conjunction with other properties of the plants (El Titi, 1972). The prey species has not been observed to have any effect on the egg-laying activity of the female (El Titi, 1974), but the species of plant and even the variety has a clear effect. According to Miesner (1975), the differences between plant species as regards oviposition are due to leaf structure, hairiness etc. He also found that midges deposited more eggs near adult aphids than larvae.

The females deposit about 100 eggs (Harris, 1973). In Uygun's (1971) experiments, the average number of eggs was only about 70, most of which were deposited during the first 2–4 days following emergence. According to El Titi (1972), the number of eggs is almost directly proportional to the aphid density.

The eggs are 0.3 mm long and 0.1 mm broad, orange in colour, smooth and shiny. At room temperature, the egg stage lasts for 2–3 days. The larvae emerge from the anterior end of the egg. Bouchard *et al.* (1981) studied the development of eggs in the laboratory at a constant temperature of 23°C when the incubation period lasted 2 days.

The newly hatched larvae are only 0.3 mm long. When fully grown, they are 2–3 mm long, elongated, narrowing at both ends.

The colour of the larvae varies from light orange to red, depending on the food source (Plate 14). There are four larval instars (Azab *et al.*, 1965).

The larvae develop only if they are able to feed on aphids. They cannot survive by feeding on scale insects and mites (Azab *et al.*, 1965), but they are polyphagous as regards aphids. Observations so far show that the larvae feed on over 60 different species of aphids, but the range of prey is apparently much wider. The prey includes all the common aphid pests encountered in greenhouses, e.g. *Myzus persicae* (Sulz.), *Aphis gossypii* Glov., *Macrosiphum rosae* L. and *M. euphorbiae* Thomas.

In nature, the larvae take 7–14 days to develop, depending on the temperature and food supply (Harris, 1973). Uygun (1971) found that development took 7 days at 15°C, 3.8 days at 21°C, and 3.0 days at 27°C. According to the laboratory studies of Bouchard *et al.* (1981), the development took 5.5 days at a constant temperature of 23°C.

Immediately after hatching, the larvae start searching for aphids on which to feed. According to Wilbert (1972), the larvae must find their first prey within a few hours. He observed that newly hatched larvae are capable of moving 63 mm without food and estimated that larvae are able to detect aphids within an area of 2.7 mm².

In contrast to most other predators of aphids, *A. aphidimyza* larvae are able to locate their prey from quite a distance– 1st instar larvae from about 3 mm. According to Wilbert (1974), larvae mainly locate their prey by olfactory means, but vision may play some part in the mechanism because larvae also react to the presence of sand grains of the same size as aphids.

Aphids make only limited attempts to escape when a larva approaches. 1st instar larvae do not produce any escape reactions in young aphids, but some of the adult aphids move away when a larva approaches (Klingauf, 1967).

Larvae usually attack aphids by biting their leg joints. The larvae then excrete a toxin which paralyzes and kills the aphid. The dead aphids are initially green but gradually turn black and desiccate. Many aphids remain attached to the leaves and hang downwards, suspended by their proboscis (Bombosch, 1958).

Mayr (1975) showed that the salivary toxin paralyzes the prey within a couple of minutes. As the aphid ceases to struggle, the larva usually bites into its thorax. The contents of the aphid are dissolved by the toxin within 10 minutes and the midge larva then sucks its prey dry.

The composition of the toxin is not known. According to Mayr (1975), a homogenate prepared from the salivary glands did not contain any proteases. It contained a phenoloxidase, which was inhibited by phenolthiourea.

Investigations in Finland have shown that the saliva contains the enzyme hyaluronidase (Laurema, Markkula & Husberg, unpublished), which occurs commonly in the toxins of many animals. The salivary gland also contains glutaminic acid in clearly higher concentrations than other tissues. It is possible that the glutaminic acid is partly responsible for the paralyzing effect of saliva.

According to Uygun (1971), the larvae require only 7 small-sized *Myzus persicae* to develop into pupae, but Nijweldt referred to other observations in which midge larvae were able to complete development if they ate 5 full-grown or 15 small-sized *M. persicae.* If there are plenty of aphids present, the midge larvae kill and eat many more than are needed for their development. Uygun (1971) found that the more and the larger the aphids available, the larger the number killed and left uneaten.

At lower temperatures, larval development takes longer and the larvae eat more aphids. Similarly, the drier the air, the greater is the number of aphids eaten (Mayr, 1975).

After the larva has grown to its full size, it crawls down the stem of the plant or falls to the ground. It then burrows down to a depth of about 3 cm and builds a cocoon. This cocoon is formed from a large number of light-coloured sticky threads and is covered by small stones, aphid skins, larval excreta etc. It is oblong, about 1.8 mm long and 0.7 mm broad. The cocoons are very difficult to find in the soil. Pupation takes place 2–4 days after the cocoon has been formed. The pupa is 1.4 mm long and 0.5 mm broad. Initially it is orange-coloured but later turns brown. At room temperature, the pupal stage lasts for 10–14 days.

In natural conditions in Finland, *A. aphidimyza* overwinters in the cocoon. The larvae diapause in the cocoons from September onwards. They pupate in spring and the adults emerge in the middle of May. In the high greenhouse temperatures, diapause starts later in autumn and finishes earlier in spring.

Under laboratory conditions, aphid midges can be reared throughout the year, without diapause, if they are kept in a 16:8 hours light:dark cycle at room temperature (El Titi, 1972). Short illumination periods and low temperature induce diapause. According to Havelka (1980), the critical daylength at 20°C for midges collected from Leningrad is 17 hours and for those from Kishinev 15.5 hours. Diapause is induced in the last larval instar 1 day before the cocoon is made and in the cocoon (Havelka, 1980).

Before emergence, the pupae come out of their cocoons and move up to the soil surface. Adults emerge head forwards from a split formed behind the middle point of the pupal body. Emergence takes 2–3 minutes. Adults fly within 10 minutes of emergence (Azab *et al.*, 1965).

2.8b **SUITABILITY OF THE APHID MIDGE AS A CONTROL AGENT** One of the most important tasks in assessing the suitability of biological control agents is to determine the most suitable developmental stage for transferring them to the plants (Figure 26).

Both midge eggs and larvae can easily be transferred to the greenhouse on the leaves on which they have been reared. However, despatch to the customer in these stages is not profitable because they are easily killed during transport, primarily as a result of desiccation and lack of food.

It is possible to introduce either adult midges or cocoons into the glasshouse. Adults have been spread either by transferring them straight to the glasshouse or by maintaining permanent mass production there. El Titi (1974) and Markkula *et al.* (1979), however, did not obtain good results when adult midges were spread to greenhouse cultures. El Titi (1974) obtained slightly more promising results by maintaining an 'open' midge culture in the greenhouse.

The best and easiest method is to transfer midge cocoons. Researchers in the Federal Republic of Germany (El Titi, 1974), the USSR (e.g. Bondarenko, 1975; Asyakin, 1977) and Finland (Markkula, 1973; Markkula & Tiittanen, 1977) have all come to this conclusion.

Mass production of cocoons In practice, great numbers of cocoons are needed for satisfactory control. The following five-stage, mass production method has been developed at the Agricultural Research Centre in Finland:

1. Capsicums are sown at 2-week intervals and cultivated in a greenhouse as food for *Myzus persicae*.
2. When the plants are 20–30 cm high they are placed in cages, 3 plants in each cage, and about 50 aphids are placed on each plant.
3. When the number of aphids has increased to about 2000 per plant, 70 female and 30 male midges are placed in each cage. After 2 days, the midges will have deposited about 3000 eggs. The plants are removed and the adult midges are killed.

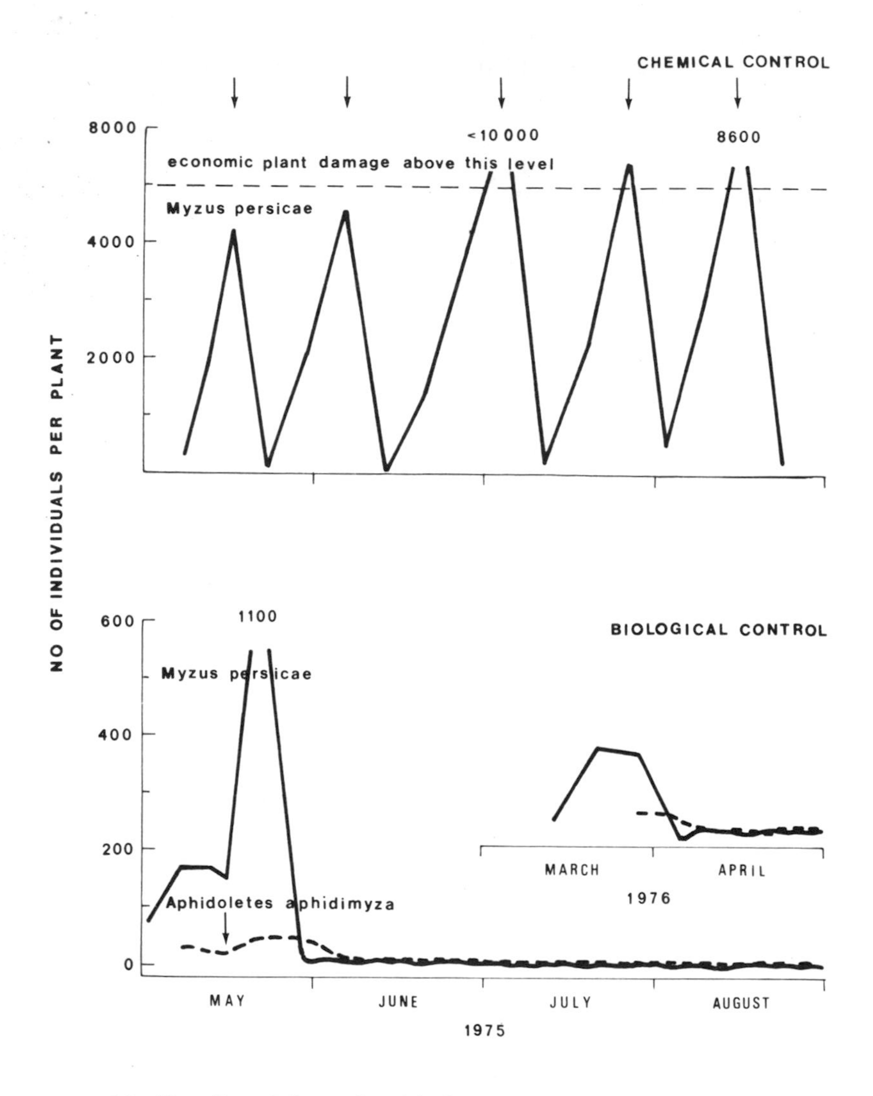

FIGURE 26 The effect of chemical and biological control of *Myzus persicae* on sweet peppers. The experiment was made in the greenhouses of the Agricultural Research Centre in Finland in 1975–76. In one greenhouse, mevinphos was used when the aphids began to damage the plants. In the other, *Aphidoletes aphidimyza* cocoons were applied at a rate of 1 cocoon for 3 aphids. The aphid midges overwintered in the greenhouse although it was not heated during mid-winter; it re-appeared on the plants the following spring, when heating began. The distribution of a single batch of aphid midge pupae into the soil gave better control than 6 treatments of mevinphos (Markkula & Tiittanen, 1982).

4. When the larvae have reached the final instar, leaves containing larvae are detached from the plants and placed on sand within small plastic containers. The containers are filled with moist sand to a height of

4 cm. On top of this is placed a sheet of nylon gauze and a thin layer of peat. 3 or 4 leaves are usually put into each container so as to give about 200 larvae per container. The larvae pupate in the peat layer.
5. The peat layer containing cocoons is sent to greenhouse growers a couple of days before the adult midges emerge.

The method of Bondarenko & Asyakin (1975) is basically the same and involves the same work stages. One essential difference is that cocoons are sieved out from the sand, their number estimated, and they are then transferred to the greenhouse without sand.

Commercial use On the basis of the studies made in the Department of Pest Investigation of the Agricultural Research Centre, a Finnish firm (Kemira Ltd) decided to commence mass production and marketing of the aphid midges in 1978.

The following instructions were drawn up for growers on the basis of research and practical experience:

> 'Peat, containing aphid midge pupae, is spread around the plants immediately after the appearance of the first aphids. One pupa per three aphids or, depending on the number of aphids present, 2–5 cocoons per m² should be applied. Treatment should be repeated after 2–4 weeks to ensure good results.'

A short description of the life cycle and behaviour of aphid midge was also added to the directions.

The control results in commercial greenhouses have been consistently good and no serious failures have occurred. Where the control has not been successful, the reason has almost invariably been because the control agents were introduced too late, when they were unable to prevent the increase of the aphids.

The main reasons for the success of the aphid midge as a biological control agent are:

1. Mass production is easy and hence economical.
2. Cocoons readily withstand transport and distribution.
3. The aphid midge forms a permanent population in the glasshouse. It can even survive the winter if the growth substrate of the plants is not changed and no harmful disinfectants used.
4. Adult midges are able to fly to aphid-infested plants even in large glasshouses.
5. The midge larvae kill and eat all the pest aphids in a greenhouse.
6. The larger the aphid population, the more aphids the midge larvae destroy.
7. The midge larvae are motile and thus are able to find new prey. Aphids do not readily escape.

The aphid midge has given control in greenhouses superior to that afforded by other predators or parasites.

REFERENCES

Asyakin, B.P. (1977) (Effectiveness of the cecidomyiid *Aphidoletes aphidimyza* Rond. (Diptera, Cecidomyiidae) in the control of aphids on vegetable crops in greenhouses.) *Zashch. Rast. Vredit. Bolez.* **53**: 121–30. (In Russian.)

Azab, A.K., Tawfik, M.F.S. & Ismail, I.I. (1965) Morphology and biology of the aphidophagous midge, *Phaenobremia aphidivora* Rübsaamen. *Bull. Soc. Ent. Egypte* **49**: 25–45.

Bombosch, S. (1958) Die Ursache eines eigenartigen Blattlaussterbens. *Z. Pflkrankh. Pflpath. Pflschutz* **65**: 694–695.

Bondarenko, N.V. (1975) Use of aphidophages for the control of aphids in hothouses. pp. 24–29. In *Proceedings of the 8th International Plant Protection Congress, Moscow* **3**.

Bondarenko, N.V. & Asyakin, B.P. (1975) (Methods for mass-rearing of *Aphidoletes aphidimyza*.) *Zashch. Rast. Vredit. Bolez.* **51** (8): 42–43. (In Russian.)

Bouchard, D., Tourner, J.C. & Paradis, R.O. (1981) Bio-écologie d'*Aphidoletes aphidimyza* (Rondani) (Diptera: Cecidomyiidae) prédatens du puceron du pommier, Aphis pomi De Gjeer (Homoptera: Aphididae). *Ann. Ent. Soc. Quebec* **26**: 119–130.

Davis, J.J. (1916) *Aphidoletes meridionalis* an important dipterous enemy of aphids. *J. Agric. Res.* **6**: 883–888.

El Titi, A. (1972) Die Verteilung der Eier von *Aphidoletes aphidimyza* (Rond.) und ihre Bedeutung für den Einsatz unter Glas. *Diss. Univ. Göttingen* 80 pp.

El Titi, A. (1974) Zur Auslösung der Eiablage bei der aphidophagen Gallmücke *Aphidoletes aphidimyza* (Diptera: Cecidomyiidae). *Entomologia Exp. Appl.* **17**: 9–21.

Harris, K.M. (1973) Aphidophagous Cecidomyiidae (Diptera): taxonomy, biology and assessments of field populations. *Bull, Ent. Res.* **63**: 305–325.

Havelka, J. (1980) Some aspects of photoperiodism of the predacious gall-midge *Aphidoletes aphidimyza* Rond. (Diptera, Cecidomyiidae). *Ent. Obozr.* **59**: 241–248.

Klingauf, F. (1967) Abwehr- und Meidereaktionen von Blattläusen (Aphididae) bein Bedrohung durch Räuber und Parasiten. *Z. Angew. Ent.* **60**: 269–317.

Markkula, M. (1973) Biological control of pests in glasshouses in Finland. *Bull. IOBC/WPRS Working Group on Integrated Control in Glasshouses, Littlehampton* **1973** (4): 7–12.

Markkula, M. & Tiittanen, K. (1977) Use of the predatory midge *Aphidoletes aphidimyza* (Rond.) (Diptera, Cecidomyiidae) against aphids in glasshouse cultures. pp. 43–44. In *Proceedings of Symposium XV of the International Congress of Entomology, Washington* USDA-ARS-NE-85.

Markkula, M. & Tiittanen, K. (1982) Possibilities of biological and integrated control of pests on vegetables. *Acta Ent. Fenn.* **40**: 15–23.

Markkula, M., Tiittanen, K., Hämäläinen, M. & Forsberg, A. (1979) The aphid midge *Aphidoletes aphidimyza* (Diptera, Cecidomyiidae) and its use in biological control of aphids. *Annls Ent. Fenn.* **45**: 89–98.

Mayr, L. (1975) Untersuchungen zur Funktion der Speicheldrüsen räuberischer Gallmückenlarven (*Aphidoletes aphidimyza* Rond.). Z. *Angew. Ent.* **77**: 270–273.

Miesner, H. (1975) Einfluss unterschiedlicher Beuteverteilung auf den Sucherfolg von *Aphidoletes aphidimyza* (Rond.). (Diptera: Cecidomyiidae). *Diss. Univ. Göttingen*, 81 pp.

Sell, P. (1976) Monogenie bei *Aphidoletes aphidimyza* (Rond.) (Diptera: Cecidomyiidae). Z. *Angew. Ent.* **82**: 58–61.

Uygun, N. (1971) Der Einfluss der Nahrungsmenge auf Fruchtbarkeit und Lebensdauer von *Aphidoletes aphidimyza* (Rondani 1847) (Diptera: Itonididae). Z. *Angew. Ent.* **69**: 234–258.

Wilbert, H. (1972) Der Einfluss der Beutedichte auf die Sterblichkeit der Larven von *Aphidoletes aphidimyza* (Rondani 1847) (Diptera: Itonididae). Z. *Angew. Ent.* **70**: 347–352.

Wilbert, H. (1974) Die Wahrnehmung von Beute durch die Eilarven von *Aphidoletes aphidimyza* (Dipt.: Cecidomyiidae). *Entomophaga* **19**: 173–181.

2.9 BIOLOGY OF GLASSHOUSE MEALYBUGS AND THEIR PREDATORS AND PARASITOIDS

M.J.W. Copland, C.C.D. Tingle, M. Saynor & A. Panis

2.9a **GENERAL BIOLOGY OF MEALYBUGS** Mealybugs (Plate 15) are small soft-bodied insects with sucking mouthparts belonging to the same order as aphids, whitefly and scale insects. They are named after the white waxy material which covers the bodies of all but the youngest nymphal instars. Most species feed on the aerial parts of plants, but some are root-feeders and others gall-producers. Mealybug feeding reduces plant vigour, causes a yellowing of the foliage, sometimes distortion and frequently defoliation of the host plant. Mealybugs produce copious quantities of honeydew. Secondary damage is caused by the growth of sooty moulds on these honeydews. The moulds cover leaves, reduce photosynthetic ability and, along with the white waxy secretions produced by the mealybugs, make the plants unsightly.

There are some 15 species recorded from glasshouses. The most common and probably the most damaging under glass, is the common or citrus mealybug, *Planococcus citri* (Risso). This species is tolerant of a broad range of environmental conditions and host plants and is found on over 25 plant families. There are two other common species. The longtailed mealybug, *Pseudococcus adonidum* (L.), has a more restricted host range and frequently conceals itself in leaf whorls. The vine mealybug (*Pseudococcus obscurus* Essig) is a cold tolerant species more common in the USA. The species are distinguished by the shape, form and length of the wax filaments, e.g. *P. longispinus* has tail filaments, which are almost as long as the body itself.

P. citri produces both sexes in appropriately equal numbers and is the only species which has been shown to mate. Other species can reproduce by parthenogenesis. The adult male is a delicate winged insect with a very short active life. Female mealybugs are wingless, ovoid insects up to 5 mm long when full grown. All species lay eggs (except *P. longispinus* which is viviparous). Eggs are small, about 0.3 mm long, and are laid within a protective mass of waxy threads. Egg-laying takes 5–10 days and the female shrinks and finally dies when the egg mass is complete. The number of eggs laid varies with temperature; *P. citri* produces less than 100 eggs above 30°C, but in excess of 400 at 18°C. The eggs hatch into a dispersive crawler stage, which searches for new feeding sites on which to settle. Following settlement, the male spins an elongate white, waxy cocoon and undergoes matamorphosis. The female nymphs have three moults and are mobile throughout their lives. Development can be completed over a wide range of temperature; *P. citri* requires about 30 days at 30°C to around 90 days at 18°C.

2.9b **CHEMICAL CONTROL OF MEALYBUGS** Mealybugs are predominantly tropical and subtropical in origin and introduction of the pest is usually via infected plant material. Thus plants should be inspected thoroughly and, if possible, quarantined before being brought into the glasshouse, ornamental collection or amenity area. The insecticides most commonly used against mealybugs include diazinon, dimethoate, formothion, malathion and nicotine. Aldicarb applied to the soil may be safely used with some biological control agents and Thripstick® (polybutenes mixed with deltamethrin) painted on to stems has also been used with some success in amenity areas. While sprays may be effective in controlling early nymphal stages of mealybugs, general chemical control is difficult for several reasons. Mealybugs tend to congregate in inaccessible places, e.g. leaf whorls, nodes, fruit calices, buds and flowers. They are protected from sprays by their waxy coat and they are often pests on ornamental plants which are sensitive to pesticides.

2.9c **BIOLOGICAL CONTROL OF MEALYBUGS** **Predators** There are many parasitic hymenopterans, predators and fungal diseases which attack mealybugs. Probably the most widely used and succesful to date is the predatory coccinellid, *Cryptolaemus montrouzieri* Mulsant (Plate 16). Insectaries in the USA produced 30 million beetles per annum primarily for release on *Citrus* (Fisher, 1963). Thc adult is a brown beetle about 4 mm long with an orange head, prothorax, wing tips and abdomen. Females mate soon after emergence and begin to lay eggs some 5 days later. They lay their eggs singly into mealybug egg masses, up to a maximum of 500 eggs, at a rate of about 10 eggs per day. The total number of eggs produced is strongly influenced by adult diet and starvation halts egg production. All stages are predatory on mealybugs. Adults and young larvae prefer host eggs and young nymphs, whilst larger larvae will consume mealybugs of any size. *C. montrouzieri* is polyphagous and will eat other homopterans, such as scales, if food is in short supply. Larvae are cannabalistic and should not be confined together without food. The larvae grow to 13 mm in length and, in the later instars, have a covering of waxy filaments, so resembling their host. At 21°C, larvae may consume over 250 2nd and 3rd instar mealybugs in completing their development. Temperature has a marked effect on the life cycle; development is complete in about 25 days at 30°C, but requires 72 days at 18°C. The adult beetles are most active in sunny conditions. Their searching behaviour is unproductive above 33°C and they become torpid below 16°C. The larvae show a similar activity pattern over this temperature range with peak useful activity at around 28°C.

C. montrouzieri has been widely used in glasshouses. Whitcomb (1940) obtained good results under favourable conditions on *Gardenia* in Massachusetts. Panis & Brun (1971) reported satisfactory control of *P. citri* on various ornamentals, provided that temperatures were above 20°C. Below this temperature, the predator's efficacy drops markedly. Because it is a predator, it is effective when hosts are plentiful, but will rarely eliminate them. Doutt (1951) found *C. montrouzieri* inadequate alone, but by supplementing with parasitic hymenopterans obtained satisfactory control. Another predatory beetle, *Nephus reunioni* Chazeau, has been used in the glasshouse in France with some success.

Parasitoids A wide range of parasitoids are found attacking mealybugs. They are generally host specific and attack a relatively narrow age range of the pest species. The most widely used parasitoid for control of *P. citri* is probably the encyrtid, *Leptomastix dactylopii* Howard (Plate 17). Techniques for mass rearing of this parasitoid are already in practice (Fisher, 1963). It is a moderate-sized chalcid, about 3 mm long and yellow brown in colour. It attacks large nymphs and adult females of *P. citri* and can lay 20 or more eggs a day at 30°C. The adults are long-lived, given food and high humidity. The life cycle is completed in 12 days at 35°C and 45 days at 18°C. It has been successfully used in many cases as a supplement to predatory beetles. In France, a combination of *Leptomastix dactylopii* and *Cryptolaemus montrouzieri* gave complete control of *Planococcus citri* on *Clivia* and crotons and reasonable control on *Pelargonium*, *Saintpaulia*, *Cattleya* and *Pilea*. *L. dactylopii* and *Nephus reunioni* gave excellent results on bromeliads. Similarly good control has been achieved in the UK on a wide range of heated ornamentals.

Mealybug populations tend to consist of many overlapping generations and a parasitoid attacking only a limited age range may take a long time to gain control. Parasitoids with different host age ranges released simultaneously are being evaluated for their control potential for *P. citri*. *Anagyrus pseudococci* (Girault) – Plate 18 – is slightly smaller than *L. dactylopii* and attacks half- to full-grown nymphs. The female is brown with distinctive white antenna and the male is generally small and black in colour. Its longevity and egg production are similar to those of *L. dactylopii* but the life cycle is always a day or two shorter. It seems to be more active under bright light conditions.

Leptomastidea abnormis (Girault) is a pale yellow brown, 2 mm long parasitoid with distinctive banded wings held aloft (Plate 19). It attacks young 2nd instar mealybugs. While the adult is shorter-lived, its egg production is slightly higher than those of the other two species. The life cycle is longer, taking 19 and 46 days at 30° and 18°C respectively. It is more sensitive to high temperature, probably due to the susceptibility of its smaller hosts.

All these parasitoids reproduce by laying an egg into an appropriately sized mealybug. The larva develops as an endoparasitoid eating out the mealybug before pupating within the skin of the mummified host. *Cryptolaemus montrouzieri* will eat newly parasitized mealybugs but appears to find them unpalatable once mummified. All three of the above parasitoids have been shown to co-exist with the predator under glasshouse conditions. During the summer months, this combination of parasitoids and predator has produced excellent control. The ladybird effectively clears areas of heavy infestation, particularly egg masses, whilst the parasitoids maintain less serious infestations at a low level with percentage parasitism up to 90% on some plants.

During the winter in glasshouses maintained at about 20°C, *C. montrouzieri* all but disappears and, whilst the parasitoids manage to maintain a small population, they seem unable to prevent pest populations from building up. Overall adequate control is achieved, but further work on regular inoculative releases is going on in an attempt to improve control during the winter.

In France, a number of parasitoids have been used for the control of other mealybug species. Complete control of *Pseudococcus obscurus* was obtained using either the parasitoid *Pseudaphycus maculipennis* (Mercet) alone on bromeliads and *Kentia* or with *C. montrouzieri* on cacti or with *N. reunioni* on ferns. Satisfactory control was obtained using *P. maculipennis* on cyclamen and with *C. montrouzieri* on *Clivia.*

Another parasitoid, *Tetracnemoidea peregrina* (Compere), has been used

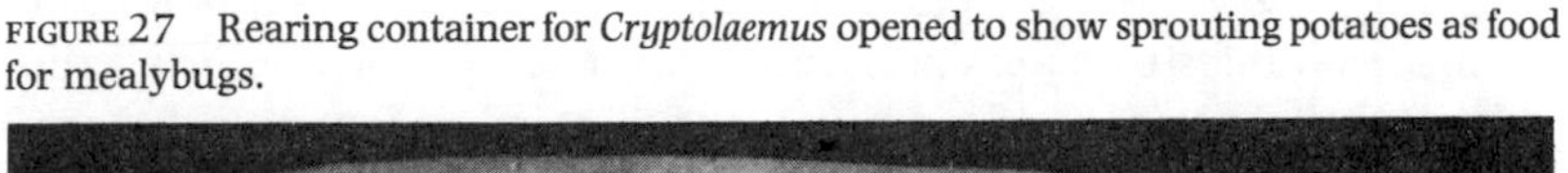

FIGURE 27 Rearing container for *Cryptolaemus* opened to show sprouting potatoes as food for mealybugs.

successfully against *P. longispinus* on ferns and *C. montrouzieri* on *Dracaena*. This combination also gave control of *Pseudococcus obscurus* on cacti.

Culture and release of natural enemies The citrus mealybug may be readily reared on potato sprouts throughout the year. At Wye, they prefer to use tubers with short sprouts raised dry rather than planted in soil at 26°C and approx 60% RH. Great care must be taken to choose sound tubers and maintain cleanliness in order to avoid infestations of mites which can rapidly destroy the culture, particularly at high relative humidities. Reasonable numbers of predators and parasitoids can be reared in plastic boxes provided with tubers infested with large numbers of mealybugs (Figure 27). Care must be taken to avoid contamination of the mealybug cultures with predators and parasitoids. One parasitoid, *Pauridia peregrina* Timberlake, has proved particularly troublesome.

Guidelines for the release of these biocontrol agents are given by Steiner & Elliot (1983). In particular, ants should be destroyed as these will protect the mealybugs for their honeydew and hamper control. Temperatures should be adjusted to favour the parasitoids and predator, i.e. 22° to 27°C for at least several hours per day. To avoid excessive heating costs, night-time temperatures could be dropped a few degrees with probably little harm to most plants. As a rough guide, predators or parasitoids should be released at a rate of 5/m^2 of infested planted area. Vents and doors should be screened to prevent escape and, during winter, some secondary glazing used to reduce heat loss and avoid trapping parasitoids in cold moisture films on the glass. Releases should be made on infested plants preferably in the early morning or evening when low light levels and temperature will prevent excessive dispersal. Plants should be monitored at least once a week by checking new growth to detect mummified mealybugs or predatory larvae. If possible, repeat releases weekly for 1 month to ensure that all stages are parasitized.

REFERENCES

Doutt, R.L. (1951) Biological control of mealybugs infesting commercial greenhouse gardenias. *J. Econ. Ent.* **44**: 37–40.

Fisher, T.W. (1963) Mass culture of *Cryptolaemus* and *Leptomastix* – natural enemies of citrus mealybug. *Bull. Calif. Agric. Exp. Stn* **797**: 38 pp.

Panis, A. & Brun, J. (1971) Essais de lutte biologique contre trois especes de 'Pseudococcidae' ('Homoptera, Coccoidae') en serres de plantes vertes. *Revue Zool. Agric. Path. Veget.* **70**: 42–47.

Steiner, M.Y. & Elliott, D.P. (1983) *Biological Pest Management for Interior Plantscapes.* Alberta Environmental Centre, Vegreville, AB. AECV83-E1.

Whitcomb, W.D. (1940) Biological control of mealybugs in greenhouses. *Bull. Mass. Agric. Exp. Stn* **375**: 22 pp.

2.10 BIOLOGY OF GLASSHOUSE SCALE INSECTS AND THEIR PARASITOIDS

M.J.W. Copland & A.G. Ibrahim

2.10a **GENERAL BIOLOGY** Glasshouse scale insects can be divided into two broad categories. The first comprise the many species of armoured scale, or Diaspidae, which, while common, are usually restricted to a limited range of hosts. The second are the soft scales, or Lecanidae, of which there are several very common and troublesome members which are highly polyphagous. These include *Coccus hesperidum* L., *Saisettia coffeae* (Walker) and *Saisettia oleae* (Bernard). Studies in Britain have concentrated on *S. coffeae* which, in our experience, has the widest host range in heated glasshouses. We have found *C. hesperidum* more common on interior plantings. These scale insects are essentially pests of the tropics and subtropics (El-Minshawy & Saad, 1976). Soft scales are round dome-shaped insects which, when full grown, are up to 5 mm in length. Young scales are usually light brown in colour becoming brown or almost black at maturity. The female scale insect produces a mass of eggs (Plate 20) over many days and then dies. In *C. hesperidum*, these hatch within her body and number from 80 to 250. In *S. coffeae*, from 500 to 2000 or more eggs are produced beneath the female's body while *S. oleae* has rather fewer eggs. The egg hatches into a dispersive crawler stage which settles onto a suitable part of the plant. Mortality at this stage may be as high as 80% and is influenced by both humidity and plant host. Soft scales feed from the phloem and may be found on the stem or on leaves where they are associated with the veins. While the plant may show little signs of damage from individual feeding, a large population will cause yellowing or defoliation. The main damaging effects are caused by the growth of sooty moulds on the copious amounts of honeydew which the scales produce. *C. hesperidum* produces considerably more honeydew than the other species. While scale insects are relatively slow-growing, they compensate for this by their large numbers of offspring. A few individuals left unchecked can therefore build into a very large population by the end of the season. We have recorded over 37 plant families attacked, ranging from ferns to orchids, and including representatives of most flowering and decorative leaf ornamentals. The family Acanthaceae seems to be the most susceptible to *S. coffeae* and also the most tolerant of high populations.

The duration of the life cycle of *S. coffeae* ranges from 95 days at 18°C to 51 days at 28°C. They are unable to develop at temperatures of 30°C or above. The scales therefore appear most troublesome in the autumn and spring or indoors at temperatures around 20°C. Temperature and host plant have an important effect on the size of the adult and, in turn, on the number of eggs which can be produced. On potato, the number of eggs may range from 2500 at 18°C to under 1000 at 28°C. On *Aphelandra*, this may be reduced 5-fold.

2.10b **CHEMICAL CONTROL** Scale insects are generally brought into the glasshouse on infected plant material. A thorough inspection before introducing new plants should therefore be made. They become the most troublesome when biological control of other pests is attempted. The insecticides commonly used are the same as those used against mealybugs and include diazinon, dimethoate, formothion, malathion and nicotine. Aldicarb applied to the soil may be safely used with some biological control agents and Thripstick® (polybutenes mixed with deltamethrin) painted on to stems has also been used with some success in amenity areas. While sprays may be effective in controlling early nymphal stages of scale insects, chemical control is difficult for several reasons. Older scales are frequently firmly attached to the plant and remain so after death. This may give a false impression of the effectiveness of chemical control. The eggs are protected from sprays by the mother's waxy body. Scales are often pests on ornamental plants which are sensitive to pesticides.

2.10c **PARASITOIDS** There are nearly 60 parasitic Hymenoptera, 14 predators and several pathogenic fungi which have been reared from *S. coffeae* alone. *Cryptolaemus montrouzieri* Mulsant, a beetle predatory on mealybugs, gives some control of soft scale in glasshouses. While there are better scale predators known, there appear to have been few studies on their effectiveness in the glasshouse. Studies in Britain have concentrated on the role which *Metaphycus helvolus* (Compere) – Plate 21 – might play in biological control. *M. helvolus* is a small encyrtid parasitoid measuring about 2 mm in length, which attacks young nymphal stages of the soft scale once they have settled. It has been used widely in the field for control of *S. oleae* on olive and citrus (Flanders, 1942). The sexes are readily distinguished, the females are yellow and the males dark. The females lay up to 5 eggs per day but, in addition, destroy nearly 4 times as many hosts by feeding on them. Despite their small size, these parasitoids are long-lived. Females have been kept alive for more than 2 months at 26°C if fed on honey solution. *M. helvolus* develops extremely quickly, the life cycle from egg to adult emergence taking about 33 days at 18°C or only 11 days at 30°C. A proportion of the eggs which are laid within hosts are destroyed by an encapsulation process. The frequency of encapsulation increases with a rise in temperature (Blumberg & De Bach, 1981). However, this is offset by an increase both in searching efficiency by the parasitoid and in egg production. We have found this parasitoid to be extremely effective during the summer although, admittedly, this coincides with a general depression of scale activity.

In France, several species of parasitoids have been used for scale control in glasshouses. Good control of *S. coffeae* has been obtained on ferns using either *Encyrtus infelix* (Embleton) or *Encyrtus lecaniorum* (Mayr) on *Peperomia* and *Dieffenbachia.* A combination of *M. helvolus* and either *E. lecaniorum or Diversinervus elegans* Silvestri gave satisfactory control on *Aphelandra* and *Aralia. M. helvolus* gave good control of *S. oleae* and *Saissetia privigna* (De Lotto) in combination with either *Metaphycus bartletti* or *Metaphycus lichtensiae* (Howard) respectively. Good or satisfactory control of *C. hesperidum* has been obtained using *M. helvolus* on *Poinsettia* and in combination with either *Metaphycus flavus* (Howard) on cacti and cymbidiums or with *D. elegans* on *Ficus* and several orchids. *Coccophagus lycimnia* Walker may also contribute to control of several soft scale species but since the male is a hyperparasite of its own on other species it should be introduced with care.

Satisfactory control of *Protopulvinaria pyriformis* Cockerell has been obtained using *M. helvolus* and *M. flavus. Scutellista cyanea* Motschulsky has given satisfactory control of *Ceratostegia floridensis* (Comstock) on *Strelitzia* but in Britain it proved to be ineffective against *S. coffeae.* There are many parasitoids known which will attack diaspid scales. Various *Aphytis* spp. have been useful against several species in the glasshouse (Steiner & Elliot, 1983).

Culture and release *M. helvolus* is easily reared on *S. coffeae.* Our rearing takes place on potato sprouts, although other members of this family are not usually heavily attacked. Good quality potato tubers are grown in soil until green leafy shoots are about 80 mm high. The tubers are then lifted, washed, and weak shoots and all roots removed. Several mature scales, with eggs near to hatching, are placed on the leaves and the crawlers allowed to settle. All rearing takes place at 26°C and 60% RH. Nymphal scales are suitable for parasitism within 2 to 3 weeks. Parasitoids are reared in plastic boxes containing tubers infested with about 100 scales per shoot. Parasitoids may be distributed within the glasshouse either within parasitized scales or released as adults. Several releases should be made at weekly intervals at a rate of about 5 parasitoids/m^2. Young parasitized scales appear slightly darker or even black, depending on the species of parasite, and emergence holes are obvious once the parasite has left. Temperatures should be adjusted to give a few hours in the range 20° to 27°C each day. In winter, secondary glazing will help to reduce heating costs and avoid losing parasites in moisture films on the glass. While *M. helvolus* does best at temperatures well over 20°C, we have found that it overwinters successfully in tropical plant collections.

REFERENCES

Blumberg, D. (1977) Encapsulation of parasitoid eggs in soft scales (Homoptera: Coccidae). *Ecol. Ent.* **2**: 185–192.

Blumberg, D. & De Bach, P. (1981) Effects of temperature and host age upon encapsulation of *Metaphycus stanleyi* and *M. helvolus* by *Coccus hesperidium*. *J. Invert. Path.* **37**: 73–79.

Flanders, S.E. (1942) *Metaphycys helvolus*, an encyrtid parasite of black scale. *J. Econ. Ent.* **35**: 690–698.

El-Minshawy, A.M. & Saad, A.H. (1976) Studies on *S. coffeae* (Walker). 1. Biology of the hemispherical scale. *Alexandria J. Agric. Res.* **24**: 515–521.

Steiner, M.Y. & Elliott, D.P. (1983) *Biological Pest Management for Interior Plantscapes*. Alberta Environmental Centre, Vegreville, AB. AECV83-E1. 30 pp.

3 BIOLOGY OF MINOR PESTS

3.1 *BRYOBIA* – THE CLOVER MITE
P. Bassett

Bryobia mites are now recognized as a complex of closely related races that are similar in appearance but which differ in life history, host plant specificity and habits. *Bryobia praetiosa*, or clover mite (Plate 22), is the species most commonly found on greenhouse crops. It is widely distributed and feeds on a variety of herbaceous plants (notably grass and clover) from which it may migrate into greenhouses.

These mites are dark reddish brown and slightly larger than the two-spotted spider mite. The body of adult females is flattened and oval in shape and bears minute, fan-shaped bristles. They are further characterized by a pair of long front legs, which are usually held outstretched in front of the body and which terminate in a pair of hooked claws. The mouthparts are similar to those of two-spotted spider mite and consist of paired stylets adapted for piercing and sucking plant tissues.

In most species of *Bryobia*, reproduction is by parthenogenesis, the males being unknown. The eggs are spherical and darker in colour than those of two-spotted spider mite. The bright red, disc-shaped larvae which hatch from these eggs have three pairs of legs. They are positively geotropic, so either migrate or drop from the egg site to herbaceous plants where they begin to feed, soon becoming dark green and almost spherical. Two nymphal stages follow before the mites reach maturity. Under greenhouse conditions, the life cycle is completed in about 4 weeks so that 5–6 generations may be expected each year.

Eggs laid in late spring hatch within a few hours, giving rise to the summer population of mites. The eggs laid by these individuals produce young stages which grow slowly during the winter to reach maturity in February or March. Some may overwinter in the greenhouse structure and emerge when the crop is planted in spring.

All stages may migrate to sheltered locations in order to moult, so very few quiescent individuals are found on herbaceous plants. The mites readily respond to disturbance by falling from their hosts, curling their legs and remaining immobile.

Where greenhouses or other buildings are surrounded by low herbage, particularly clover and grass, mites may enter through cracks to seek suitable places to moult or lay eggs. They seem particularly attracted to fresh mortar, possibly because of the high humidity. Large numbers of mites may invade houses via windows and doors but the factors which provoke these invasions are unknown.

Infestations of this pest are usually very localized on greenhouse crops. Injury to plants by *Bryobia* mites is first noticed as a winding trail of fine speckling, closely resembling damage caused by spider mites. Progressively larger areas of leaf are affected as the mite population increases and may lead to extensive damage to clovers, grasses and ornamental plants. In greenhouses, cucumbers are most commonly affected, with fruit being damaged in severe cases.

Bryobia mites are easily killed by most of the acaricides used to control two-spotted spider mites. However, where biological control programmes have led to a reduction in pesticide usage, localized outbreaks of *Bryobia* mites may cause significant damage, although serious cases on commercial cucumber crops are rare. Since attacks are usually confined to small areas of crop, overall treatment is rarely required. Localized sprays of petroleum oil are usually effective and cause minimum disruption to biological control programmes since natural enemies readily recolonise the treated areas. Observations have also suggested that the fungicide, triforine, is active against *Bryobia* mites while being relatively harmless to *Phytoseiulus* and *Encarsia.*

Where infestations from rank outside vegetation are expected, treatment of greenhouse walls may suffice.

REFERENCES

Kramer, F.W. (1956) Studies on the biology, epidemiology and control of *Bryobia praetiosa. Hofchenbr. Bayer Pflschutz.-Nachr.* **9**: 189–232.

3.2 TARSONEMID MITES
P. Bassett

The first definite record of damage by tarsonemid mites to agricultural crops was made in 1877 when *Steneotarsonemus bancrofti* was recorded as a pest of sugar cane in Queensland, Australia. Since then, many species have been recorded on crop plants, although relatively few are now regarded as being of economic importance. They occasionally cause problems in biological control programmes where severe damage may be caused to certain plants. They then assume greater importance since the acaricides necessary to control them also kill natural enemies of crop pests.

Tarsonemid mites are small (100–300 μm) and can be identified only by using a high-power microscope. As adults, they have shining oval bodies with very few setae. They are characterized by the pronounced development of apodemes – thickened rods on the ventral surface which serve as muscle attachments. In females, the hind legs are shorter and thinner than the other three pairs and terminate in two setae; one much longer than the other. They are also characterized by the presence of specialized sense organs (pseudostigmatic organs) on the dorsal surface. The males are generally smaller than females and have the hind legs modified to form a pair of claspers which are used to carry the female resting stage prior to mating.

The mouthparts consist of stout, paired palps and slender chelicerae which are unsuitable for penetrating the thick-walled and lignified tissue found in mature stems and leaves. Occasionally toxins may be injected when feeding which induce physiological changes in the host tissue.

There are 4 distinct stages in the life cycle of tarsonemid mites. Eggs are laid singly and are white, ovoid and opaque and, in some species, the surface is covered in tubercular swellings (Figure 28). The immature

FIGURE 28 Broad mite (*Polyphagotarsonemus latus*) eggs on veins of cucumber leaf.

FIGURE 29 Damage by broad mite (*Polyphagotarsonemus latus*) to cucumber fruit: normal fruit (*left*); secondary damage after control showing 'knotting of veins' (*centre*); primary damage (*right*).

larva which hatches from the egg is six-legged and is further characterized by a triangular plate-like development of the body. After this active stage, the mite becomes quiescent and the larval integument appears inflated, with the cuticle stretched. While in this state, the female is picked up by the male and held by his modified rear legs until she becomes adult. In some species, the life cycle may be completed in as little as 5 days at 20°C.

The reduced pesticide usage associated with the use of natural enemies may increase the risk of minor pests, such as tarsonemids, assuming greater importance. Two species are of most significance in biological control programmes on greenhouse crops: broad mite (*Polyphagotarsonemus latus*) and strawberry mite (*Tarsonemus pallidus*).

P. latus is widely distributed in the tropics and causes injury to crops such as cotton and tea. In temperate regions, it is becoming an increasingly common problem on greenhouse crops (Figure 29), particularly ornamentals in the UK. In 1980, *P. latus* was recorded for the first time on cucumber. Plants attacked by broad mite showed open, distorted shoots which ultimately appeared burned and shrivelled (Plate 23). Fully expanded leaves were dark green, brittle and down-curled, while fruits were russetted and cracked. Growers first attributed this damage to hormone weed-killer drift or spray contamination, but the detection of very large numbers of mites on the young leaves (200/cm^2) implicated *P. latus* in the damage (Bassett, 1981).

HV sprays containing dicofol successfully controlled this pest, although repeated applications were necessary. After acaricide application, the plants produced side shoots which showed further damage symptoms though no mites were seen on this new growth. Many small leaves were produced showing normal leaf colour, although several veins were fused and distorted, causing localized down-curling of the leaf margins. The majority of fruit produced after treatment were misshapen and showed fusion of veins in the skin although russetting was not seen.

These secondary symptoms persisted for several weeks after treatment and suggested that toxins may have been introduced by the mites when feeding earlier.

No firm conclusions were drawn on the reasons for these outbreaks of broad mite on cucumber in 1980. Biological control of red spider mite was practised on all affected nurseries. Reduced pesticide usage would favour development of broad mite. Cool, wet weather during June and July, together with growers' concern over oil prices, resulted in ventilators remaining closed while heating was switched off in most cucumber-houses. Such conditions resulted in high humidity and subsequent development of humidity-related diseases, such as *Mycosphaerella.* High humidity and moderate temperatures also favour increase of broad mite populations. However, study of meteorological data from an affected nursery showed that similar conditions had been experienced over the last 5 years. During that time, biological control had been practised continually with no symptoms of broad mite attack being seen. Furthermore, since the microclimate on the leaves and shoots of cucumber remains constant, environmental conditions within greenhouses are not thought to be a major factor. This does not exclude the possibility of the pest developing on non-crop plants outdoors.

Severe damage by broad mite has also been recorded on tomato, aubergine and sweet pepper grown in greenhouses in the UK. Symptoms of attacks were similar to those seen on cucumber with leaf margins becoming down-curled and fruit distorted or russetted. Attempts to induce symptoms on these plants by artificial infestation with mites from cucumber were unsuccessful, suggesting the presence of different strains of mite.

The wide host range of *P. latus* makes it a potentially serious pest of many greenhouse crops. The factors favouring its development are poorly understood although it is suspected that optimum conditions result from a combination of warm temperatures, high relative humidity and low light intensity. Multiplication of this species is very rapid with a generation being completed in 4 to 5 days under normal greenhouse conditions. It follows that treatment with acaricides, such as dicofol, must be repeated at short intervals to exert full control. Since this pest occurs sporadically, little experimental work has been done on the integration of suitable acaricides into integrated control programmes, although it is clear that dicofol upsets biological control when *Phytoseiulus* is being used. However, there are suggestions from the Netherlands that fenbutatin oxide (Torque®) is effective against tarsonemid mites and this acaricide has minimal effect on *Phytoseiulus.* If broad mite attacks occur late in the season, sprays of dicofol may be applied without upsetting control of spider mites since *Phytoseiulus* would be expected to have regulated spider mite populations by then.

The second economically important species, *Tarsonemus pallidus*, is variously known as strawberry, begonia or cyclamen mite and is a destructive pest of many species of ornamental plants. It is widely distributed throughout the world and sometimes occurs on plants infested with broad mite.

Attacks usually result in the distortion of young leaves which often develop a roughened, wrinkled appearance. Plants with severe damage are dwarfed and have small leaflets which fail to unfold completely. Flower buds may also become infested, causing distortion of petals so that the flowers do not open properly. Flecking and distortion of the petals may also occur on some host plants, such as *Saintpaulia*, the leaves becoming brittle and their edges curled. Severe infestations result in stunted growth.

The mites avoid light and require high relative humidity so they are found in unopened leaflets in the crown of their hosts between packed young leaves in the leaf bud. Under greenhouse conditions, all stages in the life cycle may be found throughout the year. The development is similar to broad mite, but usually takes 2 weeks from egg to adult. Eggs are usually laid in clusters within the buds.

T. pallidus is difficult to control with acaricidal sprays because the mites are well protected in the leaf buds. Care must be taken to ensure that new stock plants are not infested when brought into the nursery. If mites are found, early treatment is essential to prevent spread to other plants. Widespread use of aldicarb granules (Temik®) has helped to prevent serious outbreaks of this pest. However, where biological control programmes are used, *T. pallidus* presents serious difficulties. Heavy infestations are usually controlled by high volume sprays of endosulfan or endrin. However, both acaricides are harmful to *Phytoseiulus* while not effectively controlling two-spotted spider mite; a resurgence of this pest is, therefore, likely. Alternatively, dicofol may be used to control light infestations of *T. pallidus* which will also regulate spider mite populations, allowing natural enemies to be reintroduced later.

REFERENCES

Bassett, P. (1981) Observations on broad mite (*Polyphagotarsonemus latus*) (Acarina: Tarsonemidae) attacking cucumber. *Proceedings of the 1981 British Crop Protection Conference – Pests and Diseases* **1**: 99–103.

Crosse, J.V. & Bassett, P. (1982) Damage to tomato and aubergine by broad mite, *Polyphagotarsonemus latus* (Banks). *Pl. Pathol.* **31**: 391–393.

Hussey, N.W., Read, W.H. & Hesling, J.J. (1969) *The Pests of Protected Cultivation* Edward Arnold, London. 404 pp.

Jeppson, L.R., Keifer, H.H. & Baker, E.W. (1975) *Mites Injurious to Economic Plants* University of California Press, Berkely. pp. 285–305.

3.3 MIRID BUGS
L.R. Wardlow

Protected crops are attacked by two main species of these large (up to 7 mm long) bugs, or capsids as they are commonly called. They may enter greenhouses on plant material but more usually they fly through doorways and vents from outdoors during the summer. The common green capsid (*Lygocoris pabulinus*) is a shiny bright green insect compared with the dark-coloured tarnished plant bug (*Lygus rugulipennis*), which often has a dusky red tinge and a yellow triangle shape between the wings. Both pests can seriously damage chrysanthemums.

3.3a **COMMON GREEN CAPSID** The host range of this insect is very wide (Petherbridge & Thorpe, 1928) and includes most soft and top fruit, potatoes and many common weeds. Favourite host plants amongst flowers include rose, dahlia and chrysanthemum. There are two generations outdoors but breeding in greenhouses may be continuous. Outdoors, the pest overwinters as a cream-coloured egg inserted into the stem of its host plant, a cap at the apex of the egg usually slightly protrudes from the surface of the plant tissue. After 3 weeks, a nymph hatches from the egg to feed on the foliage. This nymph passes through five instars as it grows into an adult during the next 7 to 8 weeks. Adults lay eggs during June and July while another generation of adults appears during September and October. Eggs laid by this second generation overwinter.

3.3b **TARNISHED PLANT BUG** This bug (Plate 24) has been recorded damaging wheat, lucerne and potatoes outdoors (Southwood, 1955) and sweet pea, aster, dahlia, carnation, zinnia and chrysanthemum under glass (Hussey, Read & Hesling, 1969). Its life cycle is similar to that of the common green capsid, except that it overwinters as an adult, usually sheltering in leaf litter outdoors.

Both bugs pierce plant tissue with their mouthparts, producing a toxic saliva during the process which induces a cell reaction in the plant which forms calluses on the stems and puckering of the leaves. Both pests prefer young succulent foliage, causing distortion and blindness in young chrysanthemums. Tarnished plant bug invariably attacks opening buds, where it feeds on immature petals. Foliage attacked by the common green capsid often shows brown marks at the sites of feeding punctures and these may coalesce to form holes. Attacked petals turn brown and wilt while drops of dark brown fluid may be seen over the bloom. Tarnished plant bugs may pierce so many immature petals within the bud that the centre turns brown and the flower is eventually badly distorted. Both species are easy to find in crops but monitoring must be done carefully as, when disturbed, they readily drop off the plants or fly away.

Both pests usually occur on chrysanthemums subjected to an integrated pest control programme but, fortunately, they are unlikely to be troublesome until they enter greenhouses from outdoors in late summer. Diazinon or carbaryl HV sprays are effective and, as they do least harm to natural enemies, should be used in preference to other pesticides. Nicotine HV spray is useful where other pest/predator or parasite relationships have to be protected, although this treatment will not eliminate capsids. Nicotine fumigation is less desirable as a specific control for capsids because it is more harmful to predators and parasites, but where growers have to resort to this treatment (i.e. to control a winged aphid immigration) it is also effective against capsids.

REFERENCES

Hussey, N.W., Read, W.H. & Hesling, J.J. (1969) *The Pests of Protected Cultivation* Edward Arnold, London. 404 pp.

Petherbridge, F.R. & Thorpe, W.H. (1928) The common green capsid bug. *Ann. Appl. Biol.* **15** (3): 446–472.

Southwood, T.R.E. (1955) The nomenclature and life-cycle of the European tarnished plant bug, *Lygus rugulipennis* Poppius (Hem., Miridae). *Bull. Ent. Res.* **46**: 845–847.

3.4 BIOLOGY OF GREENHOUSE CATERPILLAR PESTS

P. Jarrett

With the increasing use of biological methods of pest control which limit the use of chemical insecticides, caterpillars are of increasing importance as pests on greenhouse crops. Also, current growing techniques can eliminate the need for steam sterilization of the soil which would otherwise kill caterpillars that pupate therein (Foster, 1979). Moths emerging from surviving pupae may produce another, larger, generation, causing more serious infestation than the original (Burges & Jarrett, 1976).

It is important to identify the caterpillars found attacking a crop as control measures differ for individual species. Also, some species are notifiable to the Ministry of Agriculture, Fisheries & Food so as to prevent their establishment in the UK.

The parent moth lays eggs on plants or even on the glass and framework of the greenhouse. On hatching, caterpillars usually eat the underside of the leaf for several days, often leaving the upper epidermis intact. Later, they eat through the whole leaf, also damaging fruits or flowers – this is often when the infestation is first discovered. Damaged foliage is usually fouled by small green faecal pellets which are often helpful in determining the site of caterpillar attack. Caterpillars normally feed at night, particularly when large, and hide during the day, or assume camouflaged postures along mid-ribs, leaf edges or petioles. When fully grown, caterpillars of noctuid moths can reach 50 mm in length and show a wide variation in colour from green to brown and black. Tortricid caterpillars do not exceed 20 mm in length and can be distinguished from young noctuids because they spin leaves together and wriggle backwards when disturbed.

3.4a **FAMILY NOCTUIDAE** *Lacanobia oleracea* (THE TOMATO MOTH) is the major caterpillar pest on tomato crops, although it is also a pest of peppers and chrysanthemums. Its general body colour varies from pale green through yellow-brown to brown and, when fully grown, it reaches a length of 45 mm. Caterpillars can be distinguished by a prominent, lateral, deep yellow line below the spiracles and well defined small, circular to oval, white spots covering the body. Other features are the white spiracles ringed by black and small hairs arising from black spots on each body segment. The moth has a wing-span of 35–45 mm, the forewings being reddish brown with nearly central brown stigmatal marks, one circular and the other kidney-shaped. The forewings have a characteristic jagged white line along the distal edge of the wings. Each female is capable of laying up to 1000 eggs in batches of between 50 and 300, which are deposited on the undersides of leaves. Eggs are hemispherical, ribbed and apple-green at first, becoming yellowish green. At 20°C, they hatch in 7 days. Newly hatched caterpillars feed gregariously on the undersides of leaves for 2 days before dispersing. Mature larvae feed not only on leaves but on stems and fruit (Figure 30) and can cause severe damage if control measures are not taken (Lloyd, 1920; Speyer & Parr, 1948). Fully grown larvae pupate in soil, crevices or under debris and will produce a second generation within the greenhouse if not prevented.

Caterpillars of *L. oleracea* can be effectively controlled by the microbial insecticide, *Bacillus thuringiensis*, applied either as an HV spray (Jarrett & Burges, 1982) or as a thermal fog (Burges & Jarrett, 1979).

Autographa gamma (THE SILVER-Y MOTH) is often found causing damage to chrysanthemums, peppers and lettuce crops under glass. The caterpillar is green, grows to 40 mm in length and is easily distinguished from the other greenhouse noctuid species in that it has only three pairs of prolegs instead of the normal five.

The moth, which has a 35–40 mm wingspan, can be readily identified by a yellow or silver Y-shaped mark centrally situated on the black to brown forewing. Up to 500 eggs can be laid by each female and these are deposited on both upper and lower leaf surfaces. Eggs are white, reticulate and hatch in 6 days at 20°C.

Caterpillars feed mainly on the foliage but will migrate upwards to feed on chrysanthemum flowers and cause economic damage. When larval growth is complete, caterpillars either spin leaves together or spin shelters in crevices of the greenhouse structure in order to protect the pupae.

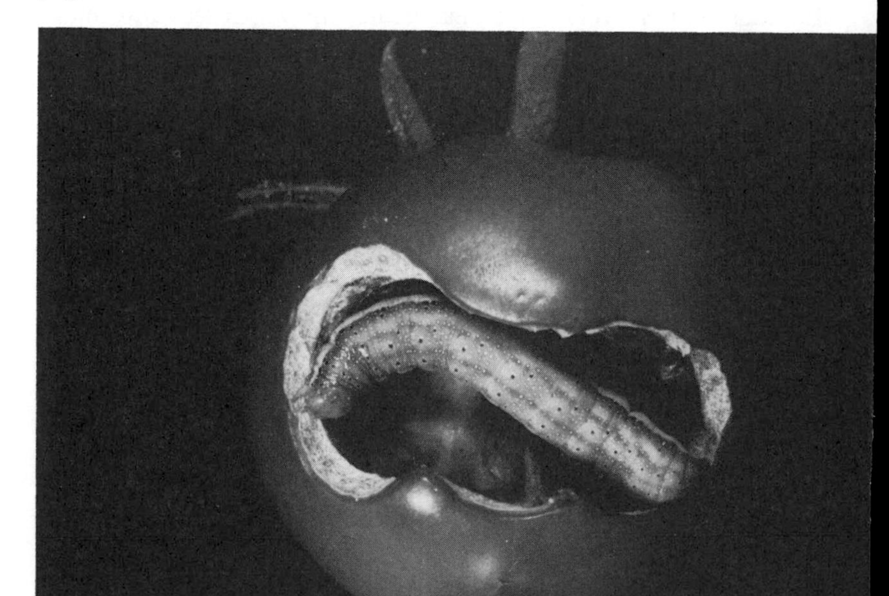

Phlogophora meticulosa (THE ANGLESHADES MOTH) The caterpillars are distinguished by a dark diagonal stripe laterally on both sides of each segment, forming a V pointing backwards when viewed from above. The fully grown caterpillar is 50 mm in length, smooth and velvety with a single, broken-white dorsal stripe. The body colour varies from green to brown and reddish brown. Caterpillars attack chrysanthemums and peppers damaging foliage, flowers and fruit.

Moths have a wingspan of 45–55 mm, the forewing is fawn, clearly patterned with V-shaped olive-green and rose-pink markings. Each female can lay up to 500 eggs which are deposited singly or in small batches, on both the upper and lower leaf surfaces.

Mamestra brassicae (THE CABBAGE MOTH) is an occasional pest on chrysanthemums and may cause severe damage to pepper crops. Caterpillars grow up to 50 mm in length and can be identified by a yellowish white band along each side, below the spiracles, and a central line speckled with white along the dorsal surface. General body colour varies from green to dark brown to almost black.

The forewings of the moth are grey-brown to black with a nearly central, distinct kidney-shaped mark bordered in white. The hindwings are light brown with a central spot. Each female lays up to 500, white, ribbed eggs in small batches on the greenhouse structure as well as on the undersides of leaves.

Naenia typica (THE GOTHIC MOTH) is occasionally found causing damage to chrysanthemums under glass. The caterpillar is distinguished by dark oblique marks down each side which produce a diamond-shaped pattern when viewed from above. Other features are a dark line along the centre and along each side. General body colour varies from grey to pale brown. Caterpillars grow to 45 mm in length before pupating in the soil.

The moth has a wingspan of 36–46 mm, the female usually larger than the male. The forewing is grey to grey-brown with two irregular, transverse, whitish brown lines and stigmata. The hindwings are grey-brown with a lighter fringe and a narrow dark line at the base. The white, ribbed and reticulate eggs are laid in clusters on the foliage.

Noctua pronuba (THE LARGE YELLOW UNDERWING) caterpillars are distinguished by a lateral, light-coloured line on either side of the body, with a rectangular dark brown or black mark above it. A white thread-like line is present on the dorsal surface but this is distinct only in the thoracic region. General body colour varies from yellowish brown to green.

FIGURE 30 Damage to tomato fruit by tomato moth (*Lacanobia oleracea*).

Caterpillars are found causing damage on chrysanthemums and peppers although they are only an occasional pest. When fully grown they are 50 mm in length.

The hindwing of the moth is distinctive in being bright orange to yellow with a broad black border. Each female can lay more than 2000 pale creamy white eggs in large batches, on the underside of leaves.

Spodoptera littoralis (THE MEDITERRANEAN CLIMBING CUTWORM) is not a native of the UK but can be transported on cuttings imported into the country. Discovery of this pest on a crop should be reported to the Ministry of Agriculture, Fisheries & Food.

Mature caterpillars can be recognized by the conspicuous black and yellow markings on each body segment. Young caterpillars are green but later become dark brown with two pairs of yellow spots on the first two segments behind the head.

The adult has brown forewings, with a complex pattern of pale lines with a clearly defined white stigmata. Eggs are laid covered in pale-brown hairs in clusters on foliage.

Heliothis armigera is occasionally transported into the UK; like *Spodoptera* it is not a native and its discovery should be reported to the Ministry of Agriculture, Fisheries & Food.

Caterpillars, which grown up to 40 mm in length, are variable in colour from black through brown to green with dark spiracles. Faint dark and light longitudinal bands run along the length of the body, a light band particularly prominent along either side.

The moth has a 40 mm wingspan, the forewings buff or tan in colour. Hind wings are light with a broad dark band. Each female lays up to 1000 creamy-white eggs which later turn brown before hatching. Eggs are small, with a diameter of 0.4 mm and are laid singly on both upper and lower leaf surfaces.

3.4b **FAMILY *TORTRICIDAE*** *Cacoecimorpha pronubana* (THE CARNATION TORTRIX MOTH) is the most common tortrix species attacking greenhouse crops, causing damage to chrysanthemums, roses and carnations. The yellowish to olive green caterpillars, have a light brown head projecting forwards with dark markings and grow up to 20 mm in length. They feed on leaves, flowers and buds, often spinning them together with silk. They pupate in a feeding site surrounded by silk.

The moth, which has orange hind wings, is small with a wingspan of 15–20 mm. The forewings are pale brown with darker brown lines. Each female can lay up to 400 eggs in sheets which are deposited both on the greenhouse structure and on the upper surface of leaves. Egg sheets are light-green in colour and covered in mucilage.

3.4c **CONTROL OF CATERPILLARS** The standard chemical control of caterpillars is usually by HV sprays of permethrin or carbaryl. A more detailed study of insecticides suitable for controlling caterpillars on greenhouse crops is discussed by Ledieu (1978). However, biological methods for the control of other pests severely limit the use of chemical insecticides. In these circumstances, good control of the most common greenhouse caterpillar pests can be obtained by applications of the insect growth regulator, diflubenzuron, or the microbial insecticide, *Bacillus thuringiensis*. Rates of application as HV sprays are 0.6 g/1 of water for diflubenzuron and for *B. thuringiensis*, 1 g/1 of water for tomatoes and 3 g/1 for chrysanthemums, peppers and lettuce.

A more detailed description of the use of *Bacillus thuringiensis* for caterpillar control is discussed in Section 4.8.

REFERENCES

Burges, H.D. & Jarrett, P. (1976) Adult behaviour and oviposition of five noctuid and tortricid moth pests and their control in glasshouses. *Bull. Ent. Res.* **66**: 501–510.

Burges, H.D. & Jarrett, P. (1979) Application and distribution of *Bacillus thuringiensis* for control of tomato moth in glasshouses. In *Proceedings of the 1979 British Crop Protection Conference – Pests and Diseases* **3**: 433–439.

Foster, G.N. (1979) Cultural factors affecting pest incidence in glasshouse tomato crops. In *Proceedings of the 1979 British Crop Protection Conference – Pests and Diseases* **3**: 441–445.

Jarrett, P. & Burges, H.D. (1982) Control of tomato moth *Lacanobia oleracea* by *Bacillus thuringiensis* on glasshouse tomatoes and the influence of larval behaviour. *Entomologia Exp. Appl.* **31**: 329–244.

Ledieu, M.S. (1978) Candidate insecticides for the control of larvae of *Mamestra brassicae* (Lepidoptera) (Noctuidae). *Ann. Appl. Biol.* **88**: 251–255.

Lloyd, L.L. (1920) The habits of the glasshouse tomato moth, *Hadena* (*Polia*) *oleracea* and its control. *Ann. Appl. Biol.* **7**: 66–102.

Speyer, E.R. & Parr, W.J. (1948) The tomato moth. *Rep. Exp. Res. Stn Cheshunt* **1947**: 41–62.

4. PRACTICAL EXPERIENCE WITH BIOLOGICAL CONTROL

4.1 WHITEFLY CONTROL BY PARASITES
N.W. Hussey

4.1a **ON CUCUMBERS** Early work on the use of *Phytoseiulus* had concentrated on cucumbers as resistance had developed rapidly and seriously on this crop – no doubt due to the higher temperatures under which it is cultivated. By 1965, Hussey, Parr & Gould had developed effective commercial methods for the use of this predator but it was immediately apparent that chemical control of whitefly, as then practised, would interfere with this potential use of biological control.

Encouraged by the apparently effective use of the parasite, *Encarsia formosa*, in the UK during the 1930s, experiments began to develop a simple, yet effective, technique for introducing it into the large houses then in use. It was recognized that the systems of culture in use before World War 2, which utilized long, low, wooden glasshouses, were regarded as economic at much lower productivity levels than would be accepted today.

Following experiments in small greenhouses at the GCRI, commercial trials were set up by advisory entomologists (ADAS) at seven different nurseries collectively occupying about 0.6 ha (Gould, 1971).

Two methods were compared: (a) introduction at the rate of 8 parasites/plant when adult whiteflies were first seen and (b) the same numbers of parasites released 2 weeks after introducing 4 adult whiteflies/plant. Both whiteflies and parasites were released from glass tubes supported on canes 1 m above the ground at the rate of 1/25 plants.

One of the natural infestations was satisfactorily controlled but the other trials did not provide effective control and chemical sprays had to be used. It proved to be very difficult to detect the first whiteflies and it was concluded that artificial whitefly infestations were likely to be more successful, despite the failures in this case thought to be due to: (a) excessive mortality in removing black scales from the leaves for tube introduction and (b) the use of only 1/25 plants, which tended to create severe, localized whitefly infestations against which it is known that *Encarsia* is less effective.

In a later series of trials based on this experience Gould *et al.* (1975), the 'classical' method, in which whitefly is deliberately introduced, was compared with the so-called 'dribble' method in which no whitefly is introduced but the parasite is released at intervals in anticipation of the pest occurring naturally. Three trials of the classical method were made, using the same introduction rates as before, after a successful smaller experiment in which each plant was infested with 2 whiteflies at planting on 11 April and the parasites introduced at a rate of 12/plant 29 days later. In the event, the maximum scale population 60 days after *Encarsia* introduction was only 19/leaf and all of these were parasitized.

Another trial, in two houses, involved planting on 15 March, infesting with 4 whitefly adults/plant 11 days later. In one house, 8 *Encarsia*/plant were introduced 14 days after the whitefly and, in the other, 4 *Encarsia*/plant were introduced 14 and 24 days after the pest. Control was achieved effectively in both houses with maximum scale populations of 130/leaf and parasitism of at least 90% within 50 days of releasing the parasite.

Variants of the 'classical' method were compared in a further trial in which whiteflies (at 4/plant) were either introduced to all plants or only to four. The latter treatment (40/plant) established a dense 'patch' of whitefly which *Encarsia* quite failed to control though parasitism in the 'classical' comparison had reached 70% within 68 days of release of the parasites without any damage through sooty moulds.

In a large commercial house containing 1500 plants, the 'classical' method was studied on cordons. The crop was planted on 25 January, 2.5 whitefly/plant were introduced on 5 March (13 scales to every fifth plant) and the parasite introductions were made differently at the east and west ends of the house. In the west, 50 parasitized scales were introduced to every fifth plant on 18 March and again on 16 April and in the east, 40 parasites were introduced on every other plant (both methods therefore used 20 parasites/plant).

Parasitism reached 50% in 4 weeks, and 90% a month later, confirming that predictable results follow the introduction of *Encarsia* where an even distribution has been assured by artificial manipulation.

It was, however, recognized that most growers would object to the principle of 'pest-in-first' and so further improvements in the 'introduction at sight' methods were sought. This was necessary in view of the difficulty of detecting small numbers of adult whiteflies. Indeed, observations at the GCRI revealed differences of as much as 6 weeks in the dates on which different observers reported whitefly following secretive introductions. There have been many attempts to improve the monitoring of low whitefly populations by yellow traps, pheromones etc. but none have suggested any reliable practical techniques. The concept of introducing parasites regularly, in small numbers, is a further attempt based on the assumption that *Encarsia* will 'search and find' more effectively than the human observer.

In 1972, trials were made at the usual rate of 8 *Encarsia*/cucumber on the day the crop was planted and repeated a fortnight later. Peak parasitism was reached in June (96 days after *Encarsia* release). However, in two other sites where the 'dribble' began 3–6 weeks after planting, the technique failed, leading to almost half the plants being infected with sooty moulds and as many as 6 application of pesticides being required to bring the situation under control The failure was caused by the development of a severe infestation of whitefly before the first parasite release. It is, of course, well known that the honeydew produced by local concentrations of adult whiteflies reduces the speed with which *Encarsia* detects its host so that biological control of dense patches of whitefly is inevitably poor.

The evidence, therefore, suggested that 'dribbling' of parasites should begin as early as possible so as to ensure that parasites are constantly available to search for the progeny of the first whiteflies to become established on the crop. To achieve this, and taking account of the all too common infestation of cucumber plants while still in the propagating phase, a revised 'dribble' programme involving the introduction of 1 parasite to every 2 plants during the last week of propagation followed by 4 introductions of 1 *Encarsia*/plant at weekly intervals was adopted. These introductions were made at a density of 1 site/50 plants, the locations being altered for each successive release.

In 1973, although all developmental stages of whitefly were present on 30% of the plants when *Encarsia* was introduced on 21 December, both host and parasite developed slowly and 5 parasite introductions had been made by the time the first black scales were observed on 29 January. However, parasitism increased steadily to 100% by May and no sooty moulds developed.

Similar results were obtained in 1974, where 50% of the plants were infested in the propagating house and black scales were not seen until 3 March after 8 introductions of *Encarsia.* When only 10% of the plants were infested before planting, 10 introductions were made before black scales appeared in late March. In both cases, excellent control had been achieved by May.

Stenseth & Aase (1983) reported 11 similar experiments comparing the 'dribble' method with releases made on the basis of the adult whitefly density at the time of the first introduction of *Encarsia.*

All the 'dribbles' started in February when the plants were already infested with between 1 and 42 whiteflies/100 upper cucumber shoots and, in each case, between 5 and 10 introductions were made before achieving successful control.

The 8 trials using 3 introductions of between 2 and 9 parasites/plant onto established whitefly infestations of up to 28 adults/100 shoots ensured control. Despite this history of success, experience in the

Netherlands was very different. Woets and van Lenteren (1976) suggested that failures were due to the fact that cucumber is a very favourable host for the whitefly in that longevity, fecundity, rate of development are greater than other hosts. Further, *Encarsia* performs less effectively on cucumber compared with tomato because (a): the form of the venation and long leaf hairs reduce walking speed by 50% and (b) the wasps pick up honeydew more easily from the long leaf hairs and so spend twice as much time preening themselves as on tomato.

However, two other factors have affected the attitude of both research workers and growers in the Netherlands. Firstly, whitefly has been effectively controlled by hydrogen cyanide while widespread use is made of organophosphorous pesticides for thrips control which rapidly upset the development of *Encarsia* populations. These circumstances have largely prevented an enthusiastic drive to develop an effective commercial method of use.

The GCRI feel that the 'pest-in-first' method is preferable on this crop, although it is interesting that the largest cucumber nursery in Europe (Humber Growers, near Hull) has developed its own production unit and achieves control by introducing large numbers of parasites regularly. The operation proves highly economic to Humber Growers as 'home' production of natural enemies is much cheaper than purchase from specialist companies.

A quite different technique for control of whiteflies on cucumbers became possible when Hall (1982) developed the fungus, *Verticillium lecanii*, as a microbial pesticide. Commercial formulations sold by Tate & Lyle Ltd of Reading contain a food substrate which permits limited saprophytic fungal growth and sporulation on the leaf surface so increasing the likelihood of infection of mobile stages of the whitefly. All stages, except the egg, are attacked – the first diseased insects usually becoming evident 6 days after application.

4.1b **ON TOMATOES** Early experiments in small greenhouses showed that *Encarsia* should be introduced onto small whitefly populations and that both pest and parasite should preferably be evenly spread through the crop. Parr *et al.* (1976) describe a series of empirical trials in commercial greenhouses which compared several different methods of achieving this objective.

The 'pest-in-first' method pioneered by the GCRI was initiated by infesting every 100th plant (3000/ha) with 10 whiteflies followed by the introduction of parasites at the rate of 150, 150 and 75 at the same sites 3, 5 and 9 weeks after infestation with the pest. On some nurseries, whiteflies appeared on the young plants during the propagation phase and *Encarsia* was then introduced at the rate of 40 parasites/100 plants. These intervals between successive parasite release were based upon the known rates of development of whitefly under the environmental conditions specified for growing tomatoes, i.e. 18°C during the day and 12°C at night. In all, 7 large-scale trials were made and in all the 'pest-in-first' method gave excellent control, the whitefly population remaining at a low level throughout the growing season with parasitism at 80% and above. Very little sooty mould developed on the leaves and none on the fruit so that the growers were completely satisfied.

The advantages of this system of release have been technically confirmed but it is important to realise the vital importance of timing. *Encarsia* prefers to attack the 3rd instar of its host and although younger stages may be attacked when the preferred stages are unavailable, they are killed. Indeed, readers should be reminded that *Encarsia* requires to partake of host protein by feeding at oviposition punctua before its own eggs can be matured. It is not unusual to find scales with 20 or more 'stabs' and such hosts rarely survive. Where *Encarsia* attacks a whitefly population with a preponderence of young stages, the effect of the consequent larval mortality may prevent the whitefly population from increasing so that establishment of the parasite is prevented. The ratios of pest and parasite used are designed to prevent this undesirable situation.

Whatever the promises of this strategic approach it is not surprising that growers find it difficult to 'take the risk' of infesting their crops with a peat which they regard as a threat to their production. In view of this practical difficulty, the so-called 'dribble' or multiple introduction method was tested, as for cucumbers. In some cases, the parasite was introduced every 2 weeks after the first whiteflies were seen, in others, introductions began 2 weeks after planting in anticipation of whitefly attack. Parasites were released onto every twenty-fifth or fiftieth plant at the rate of 1/plant (36 000/ha). These 'insurance' introductions gave unpredictable results, largely because the scale of the first natural infestation of whitefly is quite unknown. Further, it was apparent that 6 or more successive introductions were necessary so that the cost of the control could become high. If growers wait until the first whiteflies are seen they may use less parasites but, all too often, the infestation may be well established with all stages of the pest present – a situation which, all too easily, leads to excessive numbers and sooty moulds before control is achieved.

Foster & Kelly (1978) conducted classical experiments with a view to reducing the number of introductions and hence the cost of the method. They took advantage of wide differences in the density and age structure of whitefly populations during the period of parasite establishment in a range of six multispan glasshouses in Scotland in 1976. Seven trials in which 2 introductions of *Encarsia* (60 000/ha) were made when only a few whiteflies were present (though none could be detected in a 500 leaf sample) achieved good control. Where higher numbers (a maximum of

7.0) of whiteflies were present, biological control failed to prevent severe sooty moulds. The 'ceiling' for effective control by this method appeared to be 0.1 whiteflies/leaf. When 3 introductions (90 000/ha) were made on crops treated with oxamyl, at 0.1 g active ingredient/plant shortly after planting, so as to eliminate whiteflies, the pest re-infested the plants within 4–11 weeks but was completely controlled by the parasite. Again, the maximum population of whitefly which could be contained successfully proved to be 0.1 whiteflies/upper leaf at introduction. These tentative conclusions were confirmed in 6 other trials where no sooty mould developed when the initial pest population was very low. In one trial, the crop was sprayed the day following the second introduction of *Encarsia* (23 April) because of large numbers of whitefly adults. The third introduction of parasites was deferred as the grower intended to revert to chemical control. However, despite 6 further sprays a severe outbreak of sooty moulds began in late May and continued for 3 months. By this time, surviving parasites had reduced the population to 3 adults/upper leaf! From this series of trials, it appeared that effective control followed introductions at an initial whitefly density of 0.08 adults/leaf but failed at 0.14.

In Ireland, workers at Kinsealey had concluded that excessive whitefly numbers at the time of parasite introduction were the cause of failure – where less than 1 adult was found on every 5–10 plants few problems were encountered. Assuming 5 upper leaves per plant, the acceptable whitefly density in Ireland was estimated to be 0.025 adults/leaf.

Perhaps a more useful practical guide from these conclusions is that *Encarsia* should be introduced at rates well in excess of the observed density of adult whiteflies – a factor of 10 is probably reasonable to be on the safe side.

Hence, the need for careful observation by the grower is of paramount importance. Stenseth (1976) drew attention to the importance of an even age structure in the parasite population at the start of the host–parasite interaction. He studied the rate of development of *Encarsia* under fluctuating night and day temperatures. His experiments showed the parasite to have a development period of 22–30 days at varying day temperatures. At 18°C night and 24°C day, the parasite adults emerge over a period of 8 days. Assuming a life span of 4 days, oviposition is therefore assured for 12 days after introducing parasitized scales. A comparison of this data for *Encarsia* with that for whitefly shows that temperatures between 21 and 24°C provide almost identical development periods for both host and parasite. Under temperatures of 18°C at night and 24°C during the day, *Encarsia* develops more rapidly than its host, which tends to counteract fluctuations in host and parasite populations.

Stenseth's greenhouse experiments also confirmed the importance of commencing sequential parasite introductions at 14-day intervals when there was less than 1 whitefly/plant and *Encarsia* was released at the rate of 2/plant.

Stenseth also drew attention to the fact that, as whiteflies lay eggs on the youngest leaves, the age of the larvae on individual leaves is relatively uniform. This also leads to an even age distribution in *Encarsia* populations. Under the day/night temperatures specified above, non-parasitized whitefly scales will begin emerging 7 days before *Encarsia*. De-leafing may, therefore, affect the host–parasite ratio. If leaves are removed before they are 2 weeks old the parasite is favoured but if leaves 5–6 weeks old are removed the whitefly is favoured. The parasite is quite unaffected if de-leafing removes leaves more than 8 weeks old and growers should bear this in mind when they expose fruit to encourage ripening. As tomatoes are normally planted in double rows, it is a useful practice to leave foliage between the rows to counteract the almost inevitable removal of parasites from the path side of the rows. Such excised leaves can be further preserved by leaving them on the ground for a few days to allow parasite emergence. In the Netherlands, it is common practice to place these leaves on the heating pipes. In general, it is preferable to de-leaf regularly and frequently so that removal of large numbers of leaves at one time is avoided.

It is evident from the foregoing that there are two principal difficulties in ensuring effective parasite control of whitefly: (a) failures caused by the use of a standard rate of parasite introduction against an unknown whitefly population – there are large differences in the effectiveness with which growers spot the first whitefly attack – and (b) on some occasions, the parasite fails to establish rapidly despite the presence of whitefly scales. In some ways these factors are linked since whitefly is a strongly aggregative species. It was shown by van Lenteren *et al.* (1976) and Ekbom (1980) that once dense colonies of the pest become established, *Encarsia* attacks them ineffectively, being deterred by the 'rain' of honeydew. Such strong aggregation makes any assessment, based on random sampling, ineffective. Hence, ideally, a complete search is demanded though growers with more than 0.5 ha would find this impractical. However, training of staff concerned with routine culture could improve the situation and it should be remembered that, when Foster & Kelly (1978) conducted their large-scale observations, they checked up to 500 plants/house before assuming that whiteflies were absent. No doubt the use of yellow traps at a density of 500 cm^2 attractive area per plant (Webb & Smith, 1980) would help to reduce search time.

The failure of *Encarsia* to establish under some circumstances has led to much speculation and discussion. From many observations it became clear that in sunny weather the problem rarely occurred. At first, it was concluded that light intensity was the critical factor but it is now widely accepted that low temperatures in dull weather during the late winter prevent parasite dispersal.

This problem is being exacerbated by the current high fuel costs which are encouraging growers to reduce glasshouse temperatures – a trend which may increase if cultivars said to crop effectively at 18°C day and 7°C night come into commercial use.

Early work at Cambridge suggested that *Encarsia* does not fly below 21°C. However, studies in the Netherlands (van Lenteren & Hulspas-Jordan, 1983) revealed that a few parasites could fly in the greenhouse even at 13°C and that flight was very common at 17–18°C. Undoubtedly, sunlight plays an important role as the leaf surfaces will rise 3–5°C above ambient and so ensure that the threshold for dispersal is exceeded.

While attempts to find strains or new species of *Encarsia* which are more adapted to cool conditions continue, there are two alternative techniques which permit the existing strains to be used effectively.

The first is based on the 'banker' concept introduced at the GCRI; indeed, for several years past, all the greenhouses at the Institute have been protected by this technique. The strategy is to set up, under favourable conditions, an interaction between host and parasite on individual potted plants. These are then introduced into the cropping house on the assumpton that, as each plant carries abundant food for *Encarsia*, newly emergent adults will readily survive until conditions encouraging flight allow dispersal in search of whitefly on the crop.

The method is initiated by preparing 'bankers' which will be introduced into crops at the rate of 50 plants/ha. Young tomato plants (though cucumbers or even 'White Burley' tobacco could be used) are raised in an isolated glasshouse heated to a minimum of 18°C, and provided with supplementary lighting of at least 900 lux to ensure vigorous growth by extending natural daylength to 16 hours.

When these plants, in 18 cm pots, have developed about 6 leaves they are each infested with 150 mature whitefly pupae from which adults are about to emerge. 3 weeks later their progeny will have developed to young 3rd instar scales, ready for parasitism. 300 parasitized scales are then introduced to each potential 'banker'. The first black parasitized scales should appear within 2 weeks and, by the time the first adult *Encarsia* begin to emerge 3 weeks later, at least 85% of the whitefly scales should have been parasitized. The plants are then treated with a non-persistent insecticide, such as pyrethrum, which kills any unparasitized adult whiteflies which have emerged to feed and ovipost on the terminal growth. The 'bankers', each of which should carry about 10 000 parasites which will continue to emerge over the ensuing 8–10 weeks, should be introduced into the crop within a week or two of planting. In the experience of GCRI, a few whiteflies from these 'bankers' spread to infest neighbouring plants but these are rapidly attacked by the enormous local *Encarsia* population which has already been set up. Since the interval between whitefly infestation and use is 8 weeks, and the plants have to be grown to a suitable size for use, growers should plan to sow the seed 4 months before they anticipate introduction into the crop. This technique has also been tested in several commercial trials using houses up to 0.4 ha in area. It is essentially a 'do-it-yourself' method for keen growers but experience suggests that it is very effective, though little publicized, as it is not in the interests of commercial rearing companies.

Another approach to the problem of establishment of *Encarsia* is to accept that, on December-planted crops, the development of a whitefly population will occur slowly during the winter. If a chemical is used with a predictable biological effect then it should be possible to alter the age structure of the population so that carefully timed introductions of parasites can be made in late March when the number of sunshine-hours begins to increase in Western Europe. This approach was developed by Hussey (1975) with fogs of pirimiphos-methyl (Actellic®). This material was chosen because it acts as a fumigant. If the fogs are applied on still evenings (when a temperature inversion forms) and the greenhouse leakage is reduced (by thermal screens or water sprays designed to seal the glass laps) then the vapour permeates throughout the crop and persists long enough to kill all stages of the whitefly except eggs and pupae. At the prevailing cultural temperature, it is then possible to time the release of *Encarsia* to coincide with the development of 3rd instar scales suitable for parasitization 17 days later and so ensure that the host–parasite ratio is altered heavily in favour of *Encarsia*. This technique has only been attempted experimentally but the concept should be borne in mind as other fumigant pesticides with differential effects on the developmental stages of whitefly become available.

Another type of problem was encountered in the Channel Islands, where it was assumed that the techniques developed in the UK and elsewhere for the control of whitefly would be directly applicable. Despite being farther south, and so more favourably placed for sunshine, establishment of *Encarsia* proved to be extremely erratic. Intensive investigations revealed that one major problem was the use of γ-HCH placed below the plastic sheets used in peat-module culture for control of the tomato leaf-miner (*Liriomyza bryoniae*). In the late winter, with the ventilators closed for long periods, the vapour concentration in the greenhouse atmosphere increased to a level toxic to *Encarsia*.

However, another, more serious, difficulty was identified. To maximize crop production from low night-temperature fuel-saving regimes, it has become commonplace, on Guernsey, to sow the long-season tomatoes very early. This technique leads to infestation of young seedlings by whitefly still breeding on old crops of the previous season. Contamination between greenhouses is encouraged by the generally mild conditions on the islands. So, when the parasite introductions began in February and March there was, all too frequently, an excessive pest population on the crop which no amount of spraying with non-persistent pesticides could reduce to bring the host-parasite interaction back 'into balance'.

It was, therefore, believed that the critical aspect of this form of tomato culture – somewhat similar to the situation on unheated crops in the Mediterranean, was to prevent establishment of whitefly before cold winter weather prevented any further invasions from uncontrolled infestations elsewhere.

The technique used has, to date, only been evaluated in a large-scale experiment at the GCRI, Littlehampton, but it was sufficiently promising to justify trialling elsewhere. In the test, seed cv. 'Dawn' was sown on 16 October and 400 seedlings were pricked out and held at 18°C and 16.5°C night from 'pricking off' until the first flowers became visible, at which time night temperatures were reduced to 12°C. 12 days after sowing, the seedlings were protected from whitefly invasion by adhesive yellow traps suspended 25 cm above the plants at a density of 1 trap/4.5 m^2 of bench. Whiteflies were released among the plants 17, 24 and 42 days after sowing. Destructive sampling in mid-December, 52 days after sowing, revealed that the mean number of eggs and nymphs had been reduced from 49.8 ± 14.0 on unprotected control seedlings to only 12.5 ± 3.6. This reduced population, which had followed the use of traps, despite the release of hundreds of whiteflies, was to be eliminated by systemic insecticides, taking advantage of the root restriction in small pots. On the assumption that, at the propagating temperatures usually employed, whitefly eggs take 33 days to develop into pupae, oxamyl (0.4 g/plant) was applied on 8 December – 36 days after the first whiteflies were released when the seedlings were only 5 cm high. Aldicarb could also have been used although it causes an obvious marginal scorch and is not allowed on Guernsey because of the danger of contaminating the soil water. Such treatment reduced the number of scale-infested plants to only 8%, despite the fact that the released whiteflies were trapped at a density equivalent to 5/plant.

In practice, this systemic treatment would be made in mid-December when the plants are 'stood-out' so as to avoid handling treated pots. The technique should virtually eliminate all the progeny of whiteflies infesting the plants up to that time. Since there is no further introduction of plants into the house and the winter weather should preclude whitefly flight until at least March, then the normal 'dribble' technique repeated every 14 days, commencing at the end of February, should establish a viable and effective interaction to maintain a very low whitefly population on the crop throughout the growing season.

4.1c **ON ORNAMENTALS** There have been numerous attempts to use biological control on short-term crops but these have not been supported by basic research and so, not surprisingly, most of these introductions failed.

On such crops it is obvious that *Encarsia* must be established very early in the cultural cycle in order to produce a crop acceptable to the markets. An important example is *Poinsettia* which is normally sold as a potted plant. This species is only infested by whiteflies and no other pests, which simplifies the problem since no pesticides have to be applied for other purposes. With this, and many other short-term crops, whitefly infestation usually occurs in the propagating phase. This preliminary infestation may even take place in the stock beds before cuttings are taken. During this phase, stock plants are commonly protected by systemic pesticides, such as aldicarb and oxamyl. Routine use of these persistent materials has frequently led to resistance so that much reliance is now placed on persistent synthetic pyrethroids but levels of tolerance of up to 15 000× have already been demonstrated by Wardlow (ADAS, Wye).

Helgesen & Tauber (1977) showed that, at 19°C, release of *Encarsia* at a ratio of 1 parasite:30 whitefly scales reduced the whitefly population 20× by the time of harvest. These authors concluded that this effect could be sustained by releasing a mixed age-class of parasites. In practice, this would be difficult as parasites are sold only as melanized, black scales from which adult parasites emerge after a few days. Successive introductions would, therefore, have to be made to ensure that young fecund parasites are continuously present.

Obviously it is the stock plants, as well as the cuttings, that should receive parasites regularly. A rate of 1 parasite/plant/week should achieve a host–parasite ratio which ensured permanent control so long as the stock plants were only lightly infested when the treatment began. If many whiteflies were present, the population should first be reduced with insecticides.

This technique should be suitable for many species, such as fuchsia, croton, clematis (Helgesen & Tauber, 1978) on which whiteflies are a serious problem.

Verticillium lecanii (Mycotal®) is also a useful control if applications are made when it is possible to cover the plants with plastic film for 24 hours to ensure suitable conditions for spore germination.

REFERENCES

Ekbom, S.E. (1980) Some aspects of the population dynamics of *Trialeurodes vaporariorum* and *Encarsia formosa* and their importance for biological control. *Bull. IOBC/WPRS Working Group on Integrated Control in Glasshouses, Vantaq* 3 (3): 25–34.

Foster, G.N. & Kelly, A. (1978) Initial density of glasshouse whitefly in relation to the success of suppression by *Encarsia formosa* on glasshouse tomatoes. *Hort. Res.* **18**: 55–62.

Gould, H.J. (1971) Large-scale trials of an integrated programme for cucumber pests on commercial nurseries. *Pl. Pathol.* **20**: 149–156.

Gould, H.J., Parr, J., Woodville, H.C. & Simmonds, S.P. (1975) Biological control of glasshouse whitefly (*Trialeurodes vaporariorum*) on cucumbers. *Entomophaga* **20**: 285–292.

Hall, R.A. (1977) The whitefly-*Encarsia* system: a model for biological control on short-term greenhouse crops in pest management in protected culture crops. pp. 72–73. In *Proceedings of Symposium XV of the International Congress of Entomology, Washington.* USDA-ARS-NE-85.

Hall, R.A. (1982) Control of whitefly *Trialeurodes vaporariorum* and cotton aphid *Aphis gossypii* in glasshouses by two isolates of the fungus *Verticillium lecanii. Ann. Appl. Biol.* **101**: 1–11.

Helgesen, R.G. & Tauber, M. (1978) Biological control of whiteflies on Clematis in commercial greenhouse production. *N.Y. St. Flower Indust. Bull.* **93**: 4–5.

Hussey, N.W. (1975) Entomology. *Rep. Glasshouse Crops Res. Inst.* **1975**: p. 97.

Hussey, N.W. (1983) Development of a management programme to ensure whitefly free early sown tomatoes on Guernsey. *Bull. IOBC/WPRS Working Group on Integrated Control in Glasshouses, Darmstadt* **6** (3): 194–195.

Parr, W.J., Gould, H.J., Jessop, N. & Ludlam, F. (1976) Progress towards a biological control programme for glasshouse whitefly on tomatoes. *Ann. Appl. Biol.* **83**: 349–363.

Stenseth, C. (1976) Some aspects of the practical application of the parasite *Encarsia formosa* for control of *Trialeurodes vaporariorum. Bull. IOBC/WPRS Working Group on Integrated Control in Glasshouses, Antibes* **1976** (4): 104–113.

Stenseth, C. & Aase, I. (1983) Use of the parasite *Encarsia formosa* as part of a pest management on cucumbers. *Entomophaga* **28**: 17–26.

van Lenteren, J.C., Eggenkamp-Rotteveel Mansveld, M.H. & Ellenbrook, F. (1976) The host-parasite relationship between *Encarsia formosa* and *Trialeurodes vaporariorum.* V. Population dynamics in a glasshouse. *Bull. IOBC/WPRS Working Group on Integrated Control in Glasshouses, Antibes* **1976** (4): 125–137.

van Lenteren, J.C. & Hulspas-Jordaan, P.M. (1983) Influence of low temperature regimes on the capability of *Encarsia formosa* and other parasites in controlling greenhouse whitefly. *Bull. IOBC/WPRS Working Group on Integrated Control in Glasshouses, Darmstadt* **6** (3): 54–70.

Webb, R.E. & Smith, F.F. (1980) Greenhouse whitefly control of an integrated regimen based on adult trapping and nymphal parasitism. *Bull. IOBC/WPRS Working Group on Integrated Control in Glasshouses, Vantaa* **3** (3): 235–246.

Woets, J. & van Lenteren, J.C. (1976) The parasite-host relationship between *Encarsia formosa* and *Trialeurodes vaporariorum.* VI. Influence of the host plant. *Bull. IOBC/WPRS Working Group on Integrated Control in Glasshouses, Antibes* **1976** (4): 151–159.

4.2 WHITEFLY CONTROL BY FUNGI
R.A. Hall

4.2a ***VERTICILLIUM LECANII*** The use of *V. lecanii* to control whitefly was first developed on cucumbers (Hall, 1982). Success in the use of commercially supplied *Encarsia formosa* to control whitefly on this crop really depends upon infesting the crop first with the pest, then adding the parasite – a technique which, unfortunately, most growers are unwilling to use Consequently, relatively little biological control of whitefly is practised on this crop and this provided a stimulus for the development of the use of *V. lecanii.* At first, the strain of *V. lecanii* highly pathogenic to aphids (the Vertalec® strain) was tested. This gave good control only if spore suspensions were sprayed repeatedly (Kanagaratnam *et al.*, 1982); thus the ability of this strain to spread was limited (Hall, 1982). However, after a programme of strain selection in the laboratory, a fungal isolate originally obtained from whitefly was found to spread effectively after application. This isolate was used in the commercial product Mycotal®. A single spray of Mycotal® will control a whitefly population for at least 2–3 months, or longer if conditions (temperature, humidity) are optimal. Unlike *Encarsia formosa*, Mycotal® is able to control initially moderate-to-dense whitefly infestations (Hall, 1982). The first diseased scales or adults usually appear about 7 days after the spore spray but noticeable spread of disease to new, unsprayed foliage – coupled with a decrease in whitefly numbers – occurs only about 2 weeks after spraying. This considerable lag phase must be borne in mind when planning applications. In the event of a very dense infestation, it may be necessary to first reduce the population with a chemical insecticide.

V. lecanii can also be used in conjunction with *E. formosa* to control whitefly. Where conditions strongly favour the fungus, *V. lecanii* will control the pest on its own. Although the fungus will also kill a proportion of the parasite population, since the parasite is less susceptible than the whitefly, a successful co-existence between fungus and parasite can occur.

On tomatoes, the use of *E. formosa* is more successful than on cucumbers but problems sometimes arise. Parasite efficiency is greatly impaired during short dull days and certain growing techniques present special problems. For example, excessive de-leafing, demanded by the 'Guernsey arch' method, eliminates many parasites that are about to emerge as adults from pupae. Sometimes, despite the best intentions, the whitefly-parasite interaction gets out of balance in the pest's favour. Under these circumstances, a non-chemical back-up system is desirable to complement control by the parasite. *V. lecanii* is an obvious candidate for this and has now been extensively tested on the tomato crop. In these tests, whitefly control by *V. lecanii* has been very variable and, following commercial trials between 1981 and 1983, some important factors influencing the success of this fungus on tomatoes have emerged. Because the foliage on tomato plants, unlike that on cucumbers, does not provide a dense canopy within which the microclimate humidity is high, *ambient* humidity must rise to between 90% and 100% RH at night for the fungus to establish successfully. However, whether or not this is possible depends upon several factors, including the design of the greenhouse, the mode of cultivation and the seasonal conditions. In some greenhouses (e.g. very tall structures, particularly where the crop is grown in nutrient-film, rock-wool or grow-bags) conditions tend to be dry, which precludes use of *V. lecanii.* However, where tomatoes are grown in soil which is kept damp (the degree of dampness need not be excessive) the night-time relative humidity is likely to be sufficiently high. High humidity is most easily achieved in the low Dutch Venlo® type structures where the heat input and consequent drying effect of the air are minimal. In view of the foregoing, it is not surprising that the best, most reliable control has been obtained in greenhouses of this type. Other factors which tend to lower the microclimate humidity during the critical night-time period are de-leafing practices and the reluctance of growers to close the vents during periods of very hot weather because they wish to prevent greenhouse temperatures from rising above 20°C. At the other extreme, in very cold weather when night-time temperatures in a tomato crop may well be low (*ca* 12°C), it is not likely that *V. lecanii* will establish successfully.

Attempts are, at present, under way to devise methods whereby a grower will be able to determine simply and rapidly whether use of the fungus in his greenhouse is justified. A recommendation will probably never be made to replace the use of *E. formosa* by *V. lecanii* on tomatoes and their joint use is strongly urged to obtain the best possible biological control of whitefly. Joint use would have been of particular value in certain greenhouses in the UK in May/June 1983 when dull weather prevented the successful establishment of *E. formosa* on many holdings; an imbalance quickly occurred and during the subsequent very hot weather of July and August, dense whitefly populations built up which the parasite could not possibly control. By this time, *Verticillium* could not be used successfully because of the night-time venting of houses. This experience suggests that an effort should be made by growers to create conditions suitably for *V. lecanii* in the early part of the summer which would ensure the successful establishment of *E. formosa* and its effectiveness during the rest of the season.

Mycotal® has been used against whitefly on a few other protected crops. One of these is *Gerbera* (L. Wardlow, personal communication) on which *E. formosa* is not efficient. The fungus controlled whitefly populations well in most of the crop but not adjacent to heating pipes where, presumably, the air was drier and microclimate humidity lower. This example demonstrates that growers may have to become familiar with the idiosyncrasies of their greenhouses when using this type of product.

4.2b *ASCHERSONIA ALEYRODIS* Recently, some attention has been focussed on the fungus, *Aschersonia aleyrodis*. This fungus infects only whitefly scales and pupae (Ramakers, 1983); eggs are seemingly never, and adults seldom, infected. Unlike *V. lecanii*, *A. aleyrodis* does not spread to unsprayed foliage. However, under greenhouse conditions, this fungus does infect scales faster than *V. lecanii* and it may have a role in situations where the night-time humidity is high only for a short time – such as in the tomato crop. The fungus is harmless to *E. formosa* and so the two agents would complement each other. One disadvantage of this fungus is that infection rates are poor at temperatures which often prevail at night in tomato houses (<15°C). Further work is needed to assess fully the potential of this fungus in relation to its temperature needs during the critical period of high humidity when infection can occur.

REFERENCES

Hall, R.A. (1982) Control of whitefly, *Trialeurodes vaporariorum* and the cotton aphid, *Aphis gossypii*, in glasshouses by *Verticillium lecanii*. *Ann. Appl. Biol.* **101**: 1–11.

Kanagaratnam, P., Hall, R.A. & Burges, H.D. (1982) Control of glasshouse whitefly, *Trialeurodes vaporariorum* by an 'aphid' strain of the fungus, *Verticillium lecanii*. *Ann. Appl. Biol.* **100**: 213–219.

Ramakers, P.M.J. (1983) *Aschersonia aleyrodis*, a selective biological insecticide. *Bull. IOBC/WPRS Working Group on Integrated Control in Glasshouses, Darmstadt* **6** (3): 167–171.

4.3 RED SPIDER MITE CONTROL BY *PHYTOSEIULUS* IN NORTHERN EUROPE

C. Stenseth

When introduced to an isolated area such as a greenhouse the predatory mite, *Phytoseiulus persimilis*, will, if the environment is suitable, almost always eliminate a population of red spider mite (*Tetranychus urticae*). The question is how to use this predator in such a way as to prevent economic damage to the crops before the spider mite is controlled. The most important factors for successful use of *P. persimilis* are: environment, the ways in which spider mite infests the plants, the methods of predator introduction, and the use of pesticides. The latter will be treated in Section 5.

4.3a **ENVIRONMENT** A rapid build-up of spider mites during the summer is a well known problem, both in large-scale experiments and commercial practice. This is assumed to result from dry air and high temperature at the tops of the plants during hot and dry days even with open ventilators (Section 6.2). The best way to prevent this 'flare-up' of spider mite is to have a well established predator population and few spider mites by the time of year when unfavourable conditions are expected (Gould, 1970).

In areas with sub-zero outdoor temperatures, low RH may again upset the predator development during the first 3–4 weeks after planting (Stenseth, 1980).

Before *Phytoseiulus* was known to be an effective biological control agent, and before effective pesticides were available, it was customary for growers to sprinkle water on cucumber leaves to prevent the build-up of spider mite populations. Tulisalo (1977) described an automatic system which operates the sprinklers on receipt of automatically recorded information on radiation, temperature and humidity. On warm sunny afternoons, the spray operated more than 30 × and markedly reduced the rate of increase of mite populations. No doubt *Phytoseiulus* would also benefit from the higher humidities thereby created, especially in southern latitudes.

4.3b **PATTERNS OF SPIDER MITE INFESTATION** An important factor in the use of *P. persimilis* is the rate and duration of re-activation of post-diapausing spider mite females. A long duration of emergence is advantageous in maintaining a prolonged interaction of prey and predator, but the rate of emergence is an unpredictable factor which can create problems for bio-control if vast numbers of mites invade the crop before the predator is properly established (French *et al.*, 1976; Stenseth,

1980). Dry air, at this time, magnifies the problem of large populations of post-diapausing females. The size of the spider mite population in the late summer and autumn permits a prediction of the expected number of diapausing females the following spring. If vast numbers are expected, the predator should be established before the peak emergence which usually occurs 2–4 weeks after planting.

Immigrants from outside the greenhouse are the most likely cause of additional infestations during the summer. However, it is important to remember that mites are not readily dispersed by wind or carried on clothes and equipment until the population has risen to sufficient density to stimulate individuals to drop from the ends of densely infested leaves on silken threads. The degree of infestation in different greenhouses will enable growers to judge whether such dispersal is likely to be a problem on any part of his nursery. If control is good then 'invasion' through the ventilators is likely only in the hottest weather from favoured wild hosts outdoors. In France, the predator too, invades greenhouses from external sources! If the predator is not evenly established at this time spider mite populations will multiply very rapidly.

After predator introduction, an interaction period of 8–14 weeks is usual on cucumber and tomato. The spider mite and predator may not necessarily both be eradicated in the crops, but when they are reduced to low numbers, the system becomes unstable. Spider mite survivors within the greenhouse or immigrants from outside may, in these cases, cause temporary local economic damage. In the critical period 8–14 weeks after introduction, the plants should be examined and, if necessary, new predators introduced. On ornamentals, the predator will almost always eradicate both the prey and itself after about 8 weeks.

4.3c **INTRODUCTION TECHNIQUES** Several techniques of predator introduction have been described and used commercially. Without regard to the method used, too many spider mites at the time of predator introduction is the most common reason for temporary economic damage to the crop. This is especially likely to occur on tomatoes trained under 'arch' systems which results in much young foliage being invisible from below. Hence, mite populations may be much higher than expected. The problem is intensified by slow dispersal of the predator when the spider mite population is high as *Phytoseiulus* will remain and feed while food, in the form of spider mites, remains.

Introduction sites can improve the efficiency of bio-control in these cases. A leaf damage index below 0.2–0.4 at the time of predator introduction will provide the best security against temporary damage on both cucumber and tomato (Hussey *et al.*, 1965; French *et al.*, 1976). Where young cucumber plants are infested on the propagating benches, a low leaf damage index (see p. 46) at the time of introduction is especially important as young plants are more sensitive to spider mite feeding than older plants (Legowski, 1966). On chrysanthemum, introduction on every 2 m^2 of plants has given satisfactory control at an average of below 2 spider mites/leaf (Stenseth, 1981). This is a much lower rate of introduction than is practised in the UK, where 1 predator is introduced to every 10 cuttings within 2 or 3 weeks of planting. Control is then achieved by the time the 'green-bud' stage is reached so that the harvested crop is quite clean. This technique is operated between April and October but during the winter lower temperatures preclude use of the predator.

Before reading this section, it should be pointed out that successful use of the predator in a single season will virtually eliminate red spider mites from certain greenhouses so that subsequent experience with reduced mite emergence from the roof becomes quite different. Indeed, there is little doubt that the incidence of this pest has markedly declined on many large vegetable-producing nurseries following use of biological control. Growers should, therefore, carefully consider the most appropriate strategy in the light of previous experience.

Patch and uniform introduction Two principles of predator introduction have been used: 'Patch' and uniform introduction. Several variants are used for both principles.

1. 'Patch' introduction at the initial spider mite damage sites. The predator is introduced to attacked plants and later the plants are examined weekly. Further introductions are made on plants found to be infested with spider mite in the absence of the predator.

The method has been used on cucumber (Gould, 1968 & 1970; Stenseth, 1980) especially where the growing staff have been suitably trained. In many cases, workers take a real interest and play a major part in achieving effective control. The control achieved is satisfactory, but the method is laborious, because all plants have to be examined regularly, which is commercially impractical. If the plant examination is not thoroughly carried out until the predator is well established throughout the greenhouses, control may fail.

2. Where 'patch' introduction is commenced at the very first sign of spider mite damage but no further introductions are made on new outbreaks of spider mite.

With the emergence of post-diapausing females, usually spread over several weeks, this method fails to secure rapid dispersal of the predator to new infestations. This is probably the main reason for the variable control of red spider mite reported on tomato (Dixon, 1973; French *et al.*, 1976). 'Patch' introduction should only be used as described above.

3. Uniform introduction – plants are infested with spider mites just after planting and some days later (depending on leaf damage index) the

predator is introduced onto the same plants ('pest-in-first') – Plate 25.

This method has given the most reliable bio-control in several experiments on cucumber and tomato (Hussey *et al.*, 1965; Legowski, 1966; Gould, 1970; Dixon, 1973). The method may be less effective when many post-diapausing females emerge just after planting. In northern latitudes, the introduction of spider mites reared under illumination to prevent diapause may lead to re-entry into hibernation soon after release on crop plants, so that no reproduction occurs and the predator has insufficient food.

4. Uniform introduction of both spider mite and predator at the same time ('simultaneous introduction'), before the attack of spider mites becomes visible (Plate 26).

The method has been used on cucumber (Legowski, 1966; Stenseth, 1980) with the same good results as the 'pest-in-first' technique. Where large populations of diapausing females are expected to invade the greenhouse the 'simultaneous introduction' is probably preferable to 'pest-in-first'. To establish an even interaction of spider mite and predator throughout the greenhouse it is necessary that the spider mite strain used accepts the crop to which it is introduced. It is known that different strains of spider mite are adapted to different crops (Helle & Overmeer, 1973), confirming UK and Norwegian experience that spider mites bred on bean plants accept cucumber but not tomato as a new host. The vigorous growth of the cucumber plant is especially suitable for this method as the different dispersal habits of spider mite and *P. persimilis* result in a long-lasting interaction. The Koppert method of selling predators (with host mites in bottles of sawdust) is well adapted to this technique.

On chrysanthemums, there is much evidence to show that control of both aphids (Scopes, 1970a) and spider mites (Scopes, 1970b) is ensured by very early establishment of both pest and natural enemy. The pests usually occur sporadically on the cuttings so that, as in other crops, movement of parasites and predators is reduced by their initial concentration on localized infestations of the pest. One way of overcoming this difficulty is to introduce equal numbers of both pests and their natural enemies to the boxes of cuttings before planting. To date this technique has only been proven in experiments at the GCRI but, as confidence in natural enemies increases it could well be tried elsewhere.

5. Uniform introduction at the first sign of leaf damage symptoms. This technique has been used on tomato (French *et al.*, 1976), cucumber (Stenseth, 1980) and ornamentals (Stenseth, 1981). On tomato and cucumber it is essential that the predator is distributed to all attacked plants and the 'simultaneous introduction' is used on uninfested plants. If a severe spider mite infestation is expected, every third uninfested plant should be treated with 'simultaneous introduction'. Under lower spider

mite infestations it is sufficient to restrict introductions to every fifth or tenth plant.

6. Use of 'Bankers'. Another useful technique involves the use of 'bankers' to produce the large numbers of predators cheaply. For this purpose, cucumbers, planted at either end of the greenhouse, should be infested with spider mites and, when the damage index reaches 1.0, predators should be released onto the plants. A month later, thousands of predators will be available for use on the tomato crop. A succession of cucumbers raised in this manner will ensure adequate supplies of predators throughout the growing season.

Conclusions In conclusion, it must be emphasized that a uniform introduction ensures more predictable bio-control than patch introduction. The most unpredictable factor for uniform introduction is the contamination pressure of spider mites. Under high pressures, more plants must be treated than under low pressure.

The principal difficulty on tomatoes is that infestations tend to occur when the plants have reached the upper wires or, in the 'Guernsey arch' system, trained across the paths. Symptoms become visible only when populations are high so that large numbers of predators must be introduced on each infested plant according to Table 3 (p. 45).

Another difficulty on this crop is the occurrence of *Tetranychus cinnabarinus*, which produces symptoms similar to magnesium deficiency and ultimate leaf withering when only a few mites are present. In the UK, this species is referred to as the 'hypertoxic' mite. In view of the rapid plant response to damage, predators must again be released on affected plants in larger than average numbers (i.e. 50+ per plant). Fortunately, *T. cinnabarinus* is susceptible to dicofol and quinomethionate so that it may be more convenient to 'spot' treat with these acaricides.

REFERENCES

Dixon, G.M. (1973) Observations on the use of *Phytoseiulus persimilis* Athias-Henriot to control *Tetranychus urticae* (Koch) on tomatoes. *Pl. Pathol.* **22**: 134–138.

French, N., Parr, W.J., Gould, H.J., Williams, J.J. & Simmonds, S.P. (1976) Development of biological methods for the control of *Tetranychus urticae* on tomatoes using *Phytoseiulus persimilis*. *Ann. Appl. Biol.* **83**: 177–189.

Gould, H.J. (1968) Observations on the use of a predator to control red spider mite on commercial cucumber nurseries. *Pl. Pathol.* **17**: 108.

Gould, H.J. (1970) Preliminary studies of an integrated control programme for cucumber pests and an evaluation of methods of introducing *Phytoseiulus persimilis* Athias-Henriot for the control of *Tetranychus urticae* Koch. *Ann. Appl. Biol.* **66**: 503–513.

Helle, W. & Overmeer, W.P.J. (1973) Variability in tetranychid mites. *A. Rev. Ent.* **18**: 97–120.

Hussey, N.W., Parr, W.Y. & Gould, H.J. (1965) Observations on the control of *Tetranychus urticae* Koch on cucumbers by the predatory mite *Phytoseiulus riegeli* Dosse. *Entomologia Exp. Appl.* **8**: 271–281.

Legowski, T.J. (1966) Experiments on predator control of the glasshouse red spider mite on cucumbers. *Pl. Pathol.* **15**: 34–41.

Scopes, N.E.A. (1970a) Control of *Myzus persicae* on year-round chrysanthemums by introducing aphids parasitized by *Aphidius matricariae* into boxes of rooted cuttings. *Ann. Appl. Biol.* **66**: 323–327.

Scopes, N.E.A. (1970b) Prospects of biological control of pests of year-round chrysanthemums. pp. 486–487. In *Proceedings of the VIIth International Congress of Plant Protection* 21–25 September, 1970. Paris.

Stenseth, C. (1980) Investigation of uniform introduction technique for use of *Phytoseiulus persimilis* for control of *Tetranychus urticae* on cucumber. *Meld. Norg. Landbrukshøgsk.* **59** (7): pp. 12.

Stenseth, C. (1981) Biological control of *Tetranychus urticae* on greenhouse chrysanthemum using *Phytoseiulus persimilis. Gartneryrket* **71**: 270–272.

Tulisalo, U. (1977) Automated sprinkler spraying as a tool to control spider mites on glasshouse cucumber. *Bull. IOBC/WPRS Working Group on Integrated Control in Glasshouses, Antibes* **1976** (4): 177–179.

4.4 BIOLOGICAL CONTROL OF GREENHOUSE SPIDER MITES IN SOUTHERN FRANCE

M. Pralavorio, P. Millot & D. Fournier

The Mediterranean climate permits the survival of the predatory mite, *Phytoseiulus persimilis*, throughout winter under natural conditions and so this predator commonly appears spontaneously on greenhouse ornamentals and vegetables. However, such natural establishment of the predator in the relatively closed greenhouse environment is totally inadequate to control spider mite populations.

Continuously, since 1971, *P. persimilis* has been released in greenhouses in Southern France and integrated pest management programmes have been developed for both flower (particularly roses) and vegetable (cucumbers, aubergines, melons and strawberries) crops.

Problems related to the implementation of predatory control will be considered with special reference to the Mediterranean area.

4.4a **PROBLEMS RELATED TO TETRANYCHIDS** One of the first difficulties is related to the non-obligatory diapause of the two-spotted mite in the Mediterranean climate. Unlike Northern Europe, diapause induction is less affected by photoperiod than by temperature and food availability (van de Vrie *et al.*, 1972). In the Mediterranean, the latter two conditions are frequently suitable for continuous tetranychid breeding in winter. Furthermore, Rambier (1965) indicates that the geographical origin of different populations plays a role in the beginning and termination of diapause so that the number of insects in diapause varies considerably from one population to another.

Variations in the time of appearance of the first foci of tetranychid infestation in the greenhouse can be observed from one year to the next. Instead of being synchronized with the application of greenhouse heating as happens in northern countries, which allows the use of a method such as 'pest-in-first' (Anon., 1972) to prevent spring attacks, the first infestations occur between January and March, depending on the conditions in that year. However, this range can be noted within one and the same region, in the same year, depending on greenhouse location and cultural history.

As a result, crops must be monitored as soon as milder winter temperatures allow the resumption of oviposition and development of the mite population. These observations, which should take place at least twice a month from February, are essential to determine the time of introduction which is most favourable to the establishment of the predator (presence of light tetranychid infestations at different places through the greenhouse). If necessary, a release directed to isolated foci may be made prior to a generalized release. Observations should then be made on a weekly basis as soon as higher temperatures favour the rapid development of tetranychids. *P. persimilis* must be released before mites become too numerous, as rapid pest multiplication in such climates can cause severe damage.

In some cases, a release is recommended the previous autumn. This method is used on strawberries, an annual crop planted in July–August where young plants are frequently attacked by tetranychids in autumn. When the plastic mulch is laid in early December, part of the population is eliminated by removal of older leaves, but a significant inoculum sometimes persists on the young leaves to develop by February. Eradication of these tetranychids during winter can be assured by early releases from October which prevent crop damage.

4.4b PROBLEMS RELATED TO GREENHOUSE CLIMATIC CONDITIONS

Like most phytoseiid mites, *P. persimilis* can develop most effectively at high relative humidities and average temperatures of about 25°C. This species tolerates even higher temperatures up to about 35°C provided such conditions are not permanent. On the other hand, when humidity falls below 60% RH both embryonic and post-embryonic development ceases while oviposition and longevity of adults declines significantly (Pralavorio & Almaguel Rojas, 1980).

The Mediterranean climate in Southern France is characterized, especially in spring and summer, by hot and dry weather due to the mistral which keeps humidity at a very low level both day and night. Such conditions are highly beneficial to tetranychids, which become more active and migrate to the plant tips where they form typical sphere-like aggregates. Conversely, phytoseiid mites seem to totally disappear. Therefore, overhead sprays and mists should be applied to the plants to provide better conditions for predator development.

The irregular development resulting from such climatic variations causes problems in the mass production of *P. persimilis.* We had, therefore, to develop a new method for rearing in an air-conditioned room. This unit comprises a set of superimposed cylinders. Every third day, tetranychids (multiplied in a greenhouse) are introduced to a new cylinder on cut leaves. This is placed over one already containing predator mites which then move up as they become hungry. Simultaneously, the lower cylinder, from which mites have migrated, is removed. Thus, production can be rapidly fitted to demand by increasing the number of rearing units (Fournier *et al.*, In press).

4.4c PROBLEMS RELATED TO THE HETEROGENEOUS GREENHOUSE ENVIRONMENT The greenhouse is a confined space with significant differences in temperature between the roof and ground level, as well as between the walls (exposed or otherwise to the sun) and the central air space. Temperature differences may also result from different heating systems, the siting of heating pipes and the layout of ducts and ventilators. This spatial heterogeneity of the greenhouse climate induces the development of tetranychid populations in patches around the same warm places year after year.

Conversely, the greenhouse layout is unfavourable to phytoseiid mites whenever they cannot move easily from one infestation to another due to the discontinuous plant canopy (low density of foliage; wide paths). *P. persimilis* acts as an eradicating predator and dies after consuming its prey if unable to spread to other foci of infestation. Biological control fails if eradication is not almost complete, if uncontrolled foci persist long after predator release, or if further tetranychids are introduced. Conversely, reasonable spatial homogeneity of the greenhouse (dense foliage) allows predators to move from one outbreak to another to search for other prey, thus playing a protective role by covering the entire crop.

Such problems related to greenhouse heterogeneity increase in Mediterranean regions, because in a warmer climate, temperature differences between indoors and outdoors are minimal and the thermic insulation is less efficient.

4.4d **PROBLEMS RELATED TO THE USE OF PESTICIDES** Greenhouse pesticides are highly toxic to *P. persimilis* (Coulon, 1979), see Section 5. However, several major pests that are consistently encountered at certain times of the year (aphids, thrips) can be controlled only by chemicals. Other, more occasional pests (noctuids, tortricids) or crop-specific (strawberry mite, tomato russet mite) pests can also be treated only by chemicals. This is also true of fungal diseases. Furthermore, insecticides such as pyrethroids, which can be used immediately before harvest time, are very convenient for the farmer, though they do not affect tetranychids and completely eliminate predators (Croft & Whalon, 1982).

Selection of *P. persimilis* to methidathion in 1979 was successful. It was followed by selection with deltamethrin which achieved some tolerance within the strain to pyrethroids and their residues. Selection of the same strain to rotenone is currently being undertaken. By using these resistant strains, it is now possible to use some necessary chemical treatments without any significant effect on beneficial mites.

4.4e **CONCLUSIONS** Problems arising from the biological control of greenhouse tetranychids are not a serious obstacle to the use of *P. persimilis.* The major difficulties are due to unpredictable periods of tetranychid breeding in spring and also to the drier spring and summer climates which are detrimental to the predator.

Two conclusions can be drawn. Firstly, increased crop monitoring is needed to detect early infestations of the pest so that adequate predator introductions can be made. Secondly, overhead sprinkling equipment should be available to compensate for the severe dryness of the Mediterranean climate.

If the above conditions are met, the use of *P. persimilis* in Mediterranean greenhouses presents no major difficulty. The average temperature, being much higher than in northern countries, is not detrimental to the predators, which have a higher biotic potential than tetranychids at an average temperature of 30°C under high relative humidity.

REFERENCES

Anon. (1972) The biological control of cucumber pests. *Glasshouse Crops Res. Inst. Growers' Bull.* No. 1. 8 pp.

Coulon, J. (1979) Etudes de laboratoire sur la toxicité des produits phytosanitaires vis-à-vis de *Phytoseiulus persimilis* A.H. pp. 437–439. In *Proceedings of the International Symposium of the International Organization on Biological Control*, 8–12 October 1979.

Croft, B.A. & Whalon, M.E. (1982). Selective toxicity of pyrethroid insecticides to arthropod natural enemies of agricultural pests. *Entomophaga* **27** (1): 3–21.

Fournier, D., Millot, P., Pralavorio, M. (198?) Mise au point d'une nouvelle méthode d'élevage et de production de masse de *Phytoseiulus persimilis* A.H. (In press).

Pralavorio, M. & Almaguel Rojas, L. (1980) Influence de la température et de l'humidité relative sur le développement et la reproduction de *Phytoseiulus persimilis. Bull. IOBC/WPRS Working Group on Integrated Control in Glasshouses, Vantaa* **3** (3): 157–162.

Rambier, A. (1965) Etude de quelques populations de *Tetranychus urticae* Koch. *Boll. Zool. Agrar. Bachicolt.*, Series II **7**: 51–59.

van de Vrie, M., McMurtry, J.A. & Huffaker, C.B. (1972) Ecology of tetranychid mites and their natural enemies, a review. III. Biology, ecology and pest status and host-plant relations of tetranychids. *Hilgardia* **41**: 343–432.

PLATE 1 Adult whitefly.

PLATE 2 Adult *Encarsia.*

PLATE 3 Cucumber leaf showing whitefly scales parasitized by *Aschersonia aleyrodis* (orange spots) and *Encarsia formosa* (black spots).

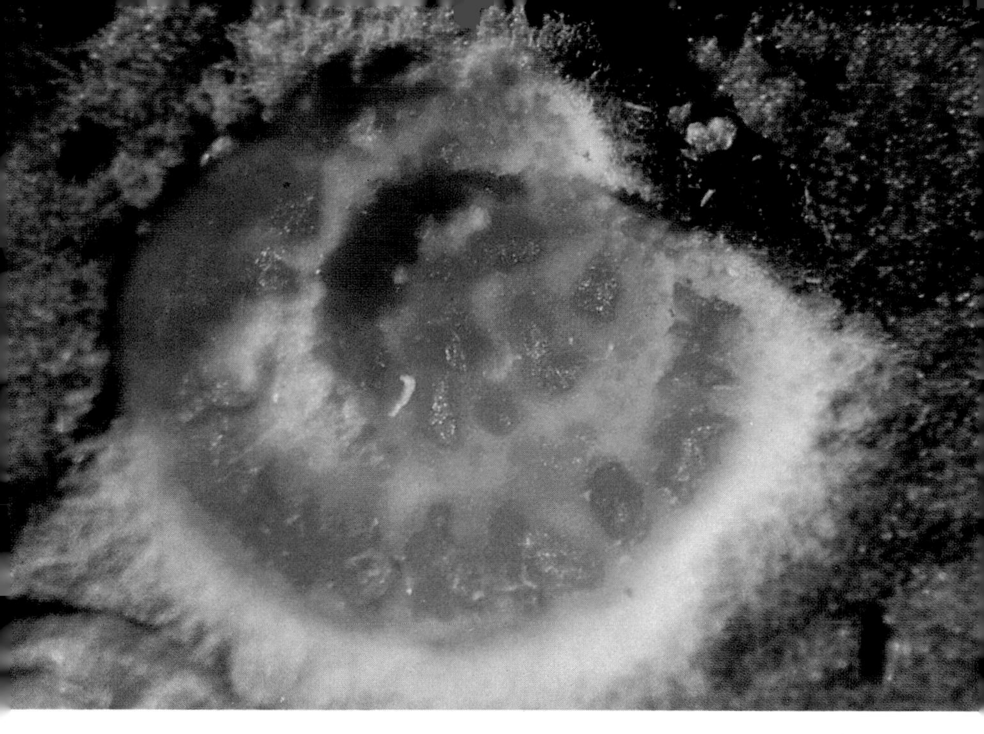

PLATE 4 Citrus whitefly infected by *Aschersonia aleyrodis*. Note bright orange of the spores.

PLATE 5 Red spider mites (green with two lateral dark spots) and the predatory mite, *Phytoseiulus persimilis* (red).

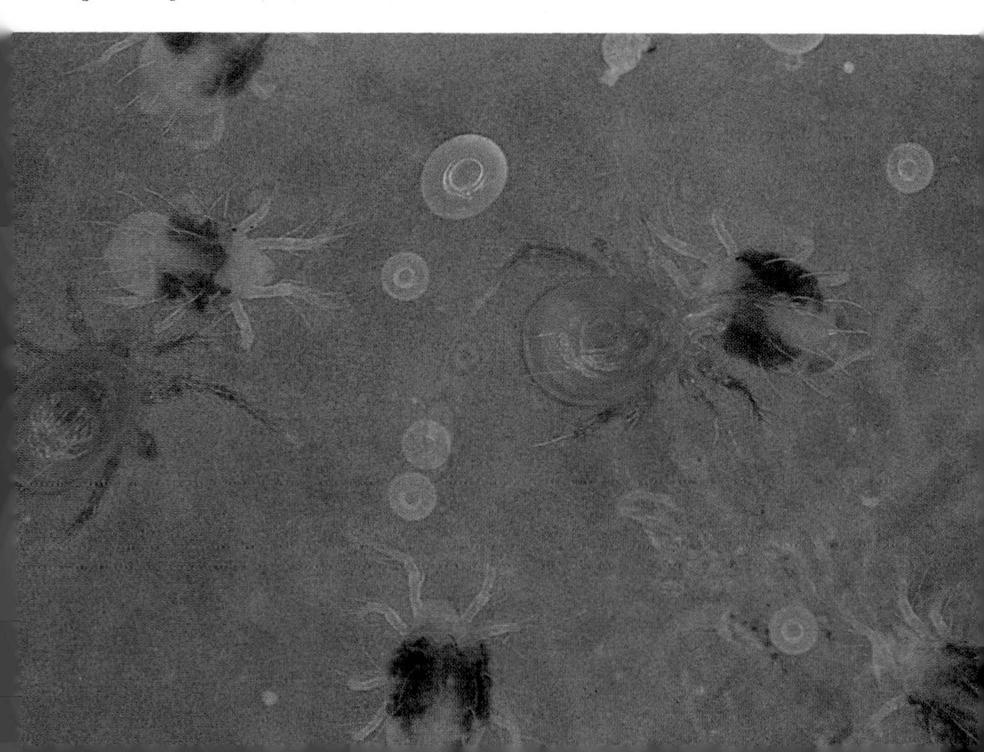

PLATE 6 Adult *Thrips nigripilosus* on chrysanthemum.

PLATE 7 The polythene bags below cucumbers to which Thripstick® is applied.

PLATE 8 Adult tomato leaf-miner, *Liriomyza bryoniae*.

PLATE 9 Adult chrysanthemum leaf-miner, *Chromatomyza syngenesiae*.

PLATE 10 Larva of *Liriomyza sativae* dissected from leaf to show ectoparasite egg of *Diglyphus pulchripes*.

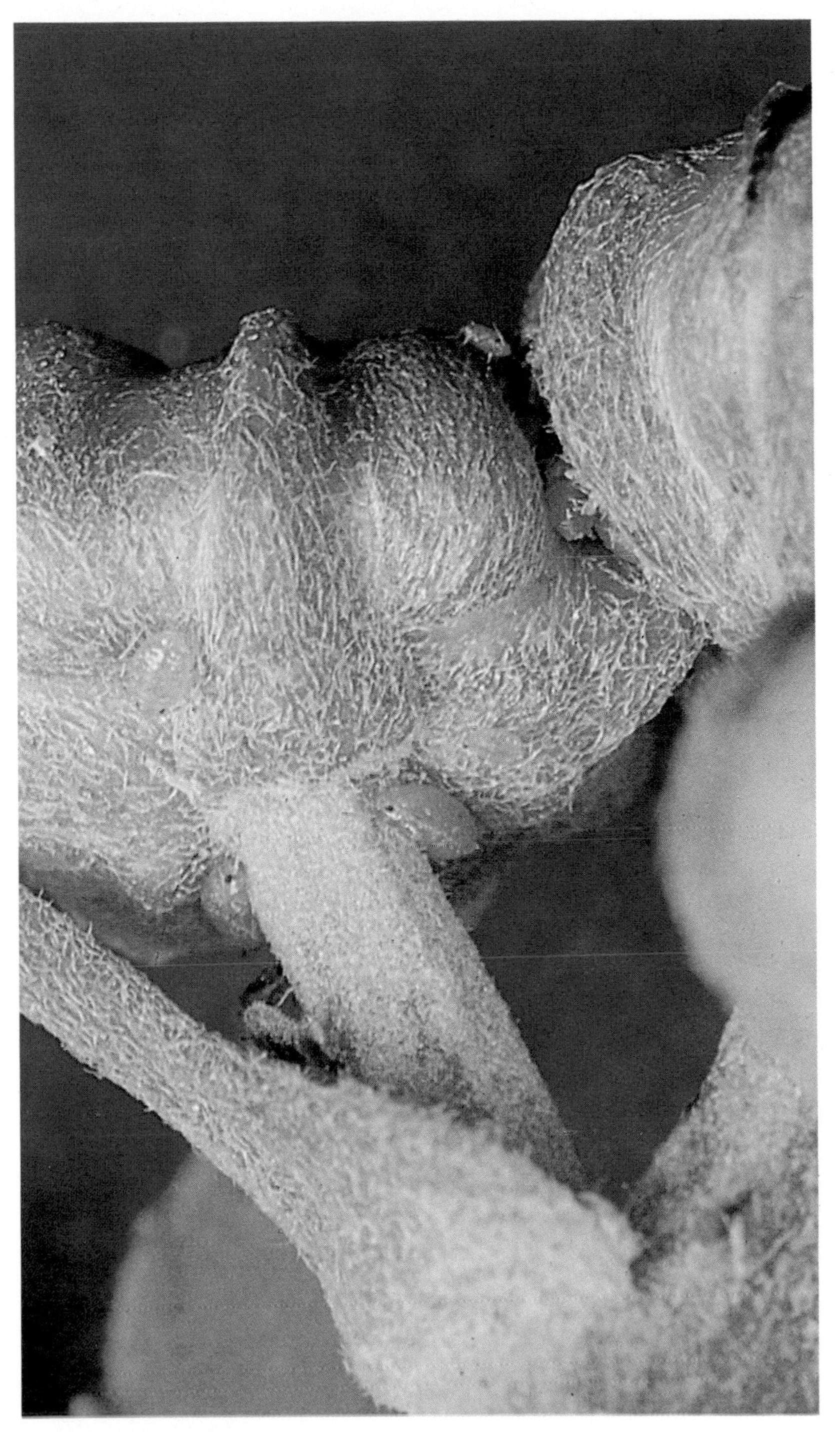

PLATE 11 *Macrosiphoniella* on chrysanthemums.

PLATE 12 *Brachycaudus* on chrysanthemums.

PLATE 13 'Mummified' aphid.

PLATE 14 Larva of midge, *Aphidoletes aphidimyza*, feeding on aphids.

PLATE 15 Colony of mealybugs.

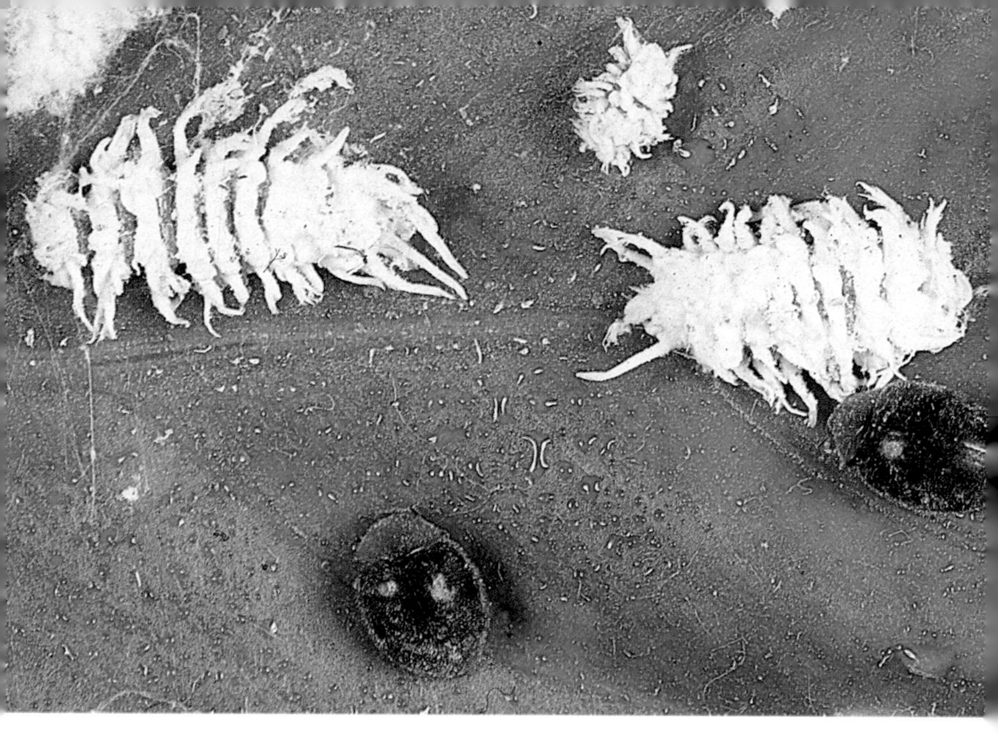

PLATE 16 Adults and larvae of *Cryptolaemus montrouzeri.*

PLATE 17 Mealybug parasite, *Leptomastix dactylopii.*

PLATE 18 Mealybug parasite, *Anagyrus pseudococci*.

PLATE 19 Mealybug parasite, *Leptomastidea abnormis*.

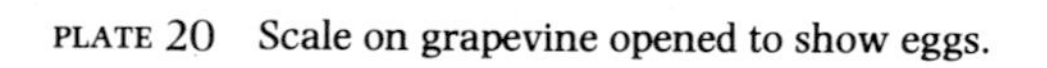

PLATE 20 Scale on grapevine opened to show eggs.

PLATE 21 Scale parasite, *Metaphycus helvolus*.

PLATE 22 Adult clover mite, *Bryobia praetiosa*, on cucumbers.

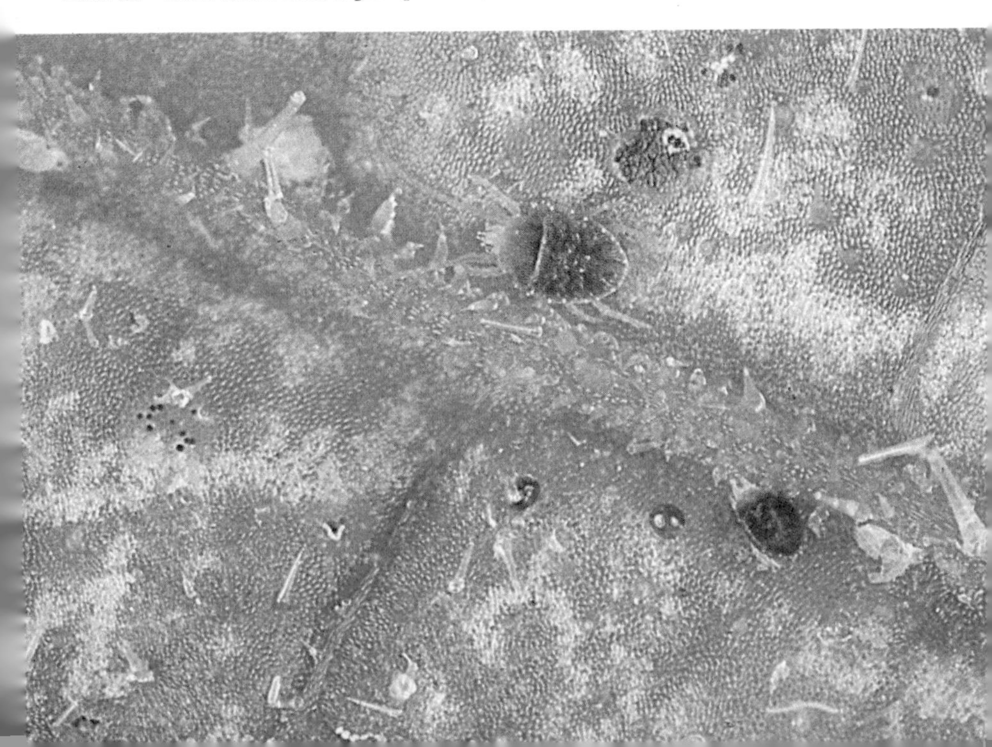

PLATE 23 Secondary symptoms of broad mite, *Polyphagotarsonemus latus.*

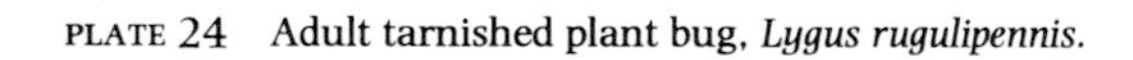

PLATE 24 Adult tarnished plant bug, *Lygus rugulipennis.*

PLATE 25 Result of delaying predator introduction until red spider mite damage is obvious.

PLATE 26 Result of introducing pest and predator together to alternate plants.

PLATE 27 Damage by 'French fly' – the mite, *Tyrophagus cucumeris.*

PLATE 28 Amenity area within public building in London.

4.5 CONTROL OF LEAF-MINERS ON CHRYSANTHEMUMS AND TOMATOES BY PARASITES

L.R. Wardlow

European entomologists have made many useful observations on the parasites of leaf-miners in recent years but little work has yet been published. Experimental results are not yet sufficiently conclusive to make firm recommendations on the commercial use of parasites. Experimental experience and the information being obtained from commercial chrysanthemum- and tomato-nurseries, is helping to establish procedures for the future use of the endoparasites, *Opius pallipes* and *Dacnusa sibirica*, as well as the ectoparasite, *Diglyphus isaea.*

4.5a **FACTORS TO BE CONSIDERED IN USING LEAF-MINER PARASITES** There are a number of considerations to be taken into account when using leaf-minor parasites:

1. Once a leaf-miner attack has developed, all stages of the pest may be found on the crop but only the larval stages are attacked by parasites. This means that, unless parasites are constantly present, especially during the early stages of the leaf-miner infestation, a proportion of the leaf-miner eggs or pupae may escape parasitism and survive to create further infestations. A single release of parasites is therefore unlikely to succeed.
2. Endoparasites live only 8–9 days producing about 90 eggs during their lifespan (Hendrikse (1980)). Ideally, this demands a parasitism rate of about 10 leaf-miner larvae/female parasite/day, but in commercial practice the rate achieved is probably much less. This work confirms that multiple releases of parasites are necessary and that these introductions should be made weekly to allow for natural mortality of the parasites.
3. Bassett (ADAS) has found that commercial supplies of parasites may be male-biased at certain times of year. We must assume that the same phenomenon occurs both in populations breeding on commercial nurseries and in natural populations on plants growing outdoors. Nurserymen should solicit the aid of entomologists in the monitoring of this problem before committing themselves to an expensive biological control programme.
4. Dull weather, cool temperatures or the use of pesticides may all affect the performance of parasites. These difficulties can also be surmounted by the use of laboratory facilities to monitor levels of parasitism once adult parasites have been released. Parasitism of larvae is best checked in a laboratory but, in the greenhouse, mortality in pupae can easily be assessed by storing samples in clear plastic containers until parasites or

leaf-miners emerge. Most scientists agree that more than 90% parasitism should be achieved rapidly and maintained for as long as possible, so that careful monitoring is necessary to achieve this objective.

5. Leaf-miners may originally infest crops as eggs or larvae in the plant material brought onto the nursery. They may also survive from one crop to the next as pupae in the soil or they may invade greenhouses from outdoors during the summer. Nurserymen must, therefore, monitor their crops regularly for signs of the pest, especially 'spotted' leaves, which denote feeding activity by the adult flies. It is usually difficult to forecast the intensity of leaf-miner attack but this evaluation may be partially achieved by regular weekly monitoring; in any event, fairly accurate assessments of the leaf-miner infestation are needed to determine the numbers of parasites to be introduced.

4.5b **CHRYSANTHEMUMS** DeLara (1981) and ADAS entomologists in England have been monitoring the performance of parasites against the chrysanthemum leaf-miner (*Chromatomyia syngenesiae*) since 1979. They have recognized that the year-round crop grows so quickly that, when young plants are attacked, the ensuing mines are quickly hidden by new foliage, making it difficult for nurserymen to monitor the pest. Infestations are frequently not discovered until a second generation attack appears on the upper leaves of mature plants.

ADAS entomologists recommend that 3 adult parasites are released per 1000 plants in the 1st week after planting each batch of young chrysanthemums in order to combat any early leaf-miner attack. A second release at the same rate is then made in the 6th week after planting to combat the progeny of any survivors. These rates of introduction (1350 parasites/ha) should be capable of controlling an infestation of 1 mine to every 6 plants, a level much greater than is normally seen on young plants. This system failed at only 1 of 15 sites monitored during 1979–82, although, at some other sites, control was variable and necessitated extra parasite introductions or partial chemical control to redress imbalances. DeLara (1981) introduced 1250 parasites/ha on 3 fortnightly occasions to crops which he monitored and found that they were successful if the first introduction was made as soon as the first adult feeding punctures were seen. Failures occurred when parasites were introduced too late, when even massive introductions of 20 000 parasites/ha were subsequently unable to redress the imbalance.

In most chrysanthemum observations, natural populations of the ectoparasite, *Diglyphus isaea*, entered the greenhouses from outdoors during June/July and helped to control leaf-miners, even by attacking larvae already parasitized by endoparasites. However, in 1982, natural immigrations of *D. isaea* were low in South-East England and commercial supplies did not improve the situation. Nurserymen were, therefore, forced to rely solely on control by endoparasites which was fortunately successful. This poor performance of *D. isaea* in that year prompts many questions and throws some doubt on its reliability for integrated pest management in Britain.

4.5c **TOMATOES** Pupae of the tomato leaf-miner (*Liriomyza bryoniae*) rest on, or in, the ground until the next generation emerges. Work by Bassett and Ledieu (GCRI) has shown that these flies may take several weeks to emerge from pupae, especially after a period of low temperature. The age structure of the subsequent leaf-miner infestation on the crop therefore depends on the rate of this emergence and must be considered when introducing parasites. The use of parasites is NOT recommended on seedling tomatoes since leaf-miner larvae (even though parasitized) may tunnel from the cotyledon into the stem, so killing the plant. This is not the case on mature plants, however, which can withstand a high level of leaf-mining. It was found that yield was not affected until there were more than 15 mines/leaf adjacent to a truss of swelling fruit. In practice, most nurserymen are alarmed by the cosmetic appearance of the crop, even when there are as few as 3 or 4 mines/leaf, so that an infestation would not normally be allowed to develop to the stage where yield is affected.

Experimental results of work on parasites in the Netherlands has been published by Zucchi & van Lenteren (1978) and Woets & van der Linden (1982 & 1983) and in England by DeLara (1981). In recent years, unpublished observations have also been made by ADAS and GCRI entomologists in England. Most workers agree that parasites should be introduced at the first signs of adult leaf-miner feeding punctures (seedlings excluded) and that several introductions of parasites are necessary. Woets & van der Linden (1983) calculated that the parasite population needs to constitute 3% of the number of larvae of the preceding leaf-miner generation in order to succeed.

In the Dutch observations, *Opius pallipes* was considered a better parasite than *Dacnusa sibirica*, though either or both successfully controlled leaf-miner if they were present at the commencement of the leaf-miner attack. In ADAS observations in the UK, *Dacnusa sibirica* was effective throughout the year but, in mid-summer, *Diglyphus isaea* became the dominant parasite, attacking leaf-miner larvae already parasitized by *Dacnusa sibirica* as well as healthy larvae. In South-East England, leaf-miner infestations that reached a peak of 3–7 mines/leaf in mid-summer were reduced by parasites to as low as 1 mine/10 leaves by September.

Recent experimental work suffers from the problem that no one has yet used what now appears to be the most effective introduction levels of parasites to control the first two generations of leaf-miner. It is believed to be necessary to introduce parasites at the rate of 1 adult for every 10 *new* mines seen on the crop each week during the first 6 weeks of an infestation. Although monitoring the pest to determine the necessary introduction levels was complicated and time-consuming, this approach was effective and achieved 72–80% parasitism in the second generation attack. However, it was evident that attacks decline only when parasitism exceeds 90% and this did not occur until later in the season. Nurserymen should realize that even if 90% parasitism is achieved in the first generation of a leaf-miner infestation, the population can still increase dramatically from 1 mine/100 plants to almost 1 mine/plant in the next generation. Even this dramatic increase is deceptively innocuous and it may take two more generations for the nurseryman to become concerned.

Due to the current high cost of leaf-miner parasites, most nurserymen are unlikely to spend more than the cost of about 4500 parasites/ha over the first 6 weeks of a leaf-miner attack. This 'ration' of parasites could be used up in 2 weeks in an attack causing a weekly population increase of 1 *new* mine/plant, but would be sufficient for 6 weekly introductions so long as there was no more than 1 *new* mine appearing on every 3 plants in any week. At current costs, if the intensity of early leaf-miner attack consistently approaches a weekly increase of 1 *new* mine to every 3 plants, the cost of biological control would probably not be economically justified.

4.5d **ANOTHER METHOD OF INTEGRATED CONTROL OF *LIRIOMYZA BRYONIAE*** Ledieu and Bassett have both successfully used a preparation of polybutenes and deltamethrin (Thripstick®) sprayed onto plastic bags and floor coverings in tomato-houses to control leaf-miners. Pupae falling to the floor stick to the preparation and adult leaf-miners are killed on emergence, while biological control methods for two-spotted spider mite or whitefly are not affected. Leaf mines were reduced from 4.7/leaf to 1.2/leaf by this method, which seems promising if it can be integrated with *Diglyphus isaea*, which does not pupate on the ground.

4.5e **MONITORING LEAF-MINER INFESTATION AND PARASITISM** No fewer than 10 plants must be chosen in each greenhouse for continual monitoring through the season. On each plant, an upper, middle and lower leaf must be clearly marked with different colours of brightly coloured adhesive tape. Each week, both the number of leaflets showing feeding marks and the number of mines per leaf should be recorded. As the plant grows and lower leaves are removed, a replacement upper leaf must be selected, in which event the previous upper and middle leaves should become the middle and lower leaves. The weekly level of adult activity (leaflets affected by feeding marks) and increases in mining, i.e. new mines, can be easily calculated from the weekly records.

Parasitism can be checked by removing leaflets with half-grown mines (i.e. containing larvae) for dissection in a laboratory and by the collection of pupae falling into trays placed beneath the plant. Pupae can be placed in clear plastic boxes to see if leaf-miners or parasites emerge.

REFERENCES

DeLara, M. (1981) Development of biological methods of pest control in the United Kingdom glasshouse industry. In *Proceedings of the British Insecticide & Fungicide Conference (1981)* **3**: 599–607.

Hendrikse, A. (1980) A method for rearing two braconid parasites (*Dacnusa sibirica* and *Opius pallipes*) of the tomato leaf miner (*Liriomyza bryoniae*). *Meded. Rijksfac. Landb. Gent* **45**: 563–571.

Woets, J. & van der Linden, A. (1982) On the occurrence of *Opius pallipes*, Wesmael and *Dacnusa sibirica*, Telenga (Braconidae) in cases of natural control of the tomato leaf miner (*Liriomyza bryoniae*, (Agromyzidae)) in some large greenhouses in the Netherlands. *Meded. Rijksfac. Landb. Gent* **47**: 533–537.

Woets, J. & van der Linden, A. (1983) Observations on *Opius pallipes* Wesmael (Hym. Braconidae) as a potential candidate for biological control of the tomato leaf miner, *Liriomyza bryoniae* Kalt (Dipt. Agromyzidae) in Dutch greenhouse tomatoes. *Bull. IOBC/WPRS Working Group on Integrated Control in Glasshouses, Darmstadt* **6** (3): 134–139.

Zucchi, R. & van Lenteren, J.C. (1978) Biological characteristics of *Opius pallipes*, Wesmael (Hymenoptera: Braconidae), parasite of the tomato leaf miner, *Liriomyza bryoniae*, Kalt. *Meded. Rijksfac. Landb. Gent* **43**: 455–461.

4.6 APHID CONTROL BY PARASITES
I.J. Wyatt

The parasites of aphids show many qualities which would recommend them as potential biological control agents, since they are highly adapted to their function as parasites. However, it is generally accepted that, ecologically, the most successful parasites are those which can exploit their host without leading to its extinction. It may be for this very reason that hymenopterous parasites of aphids have not shown the promise that one would expect and have not yet been successfully exploited in greenhouses on a commercial scale. We will now examine some of their advantages and disadvantages.

4.6a **NATURAL INTRODUCTIONS** Aphid parasites have occasionally been observed occurring naturally among greenhouse aphids and achieving effective control. *Myzus persicae* is often attacked by *Aphidius matricariae* on chrysanthemums, but control is not usually achieved until the plants are in bloom and the foliage disfigured by numerous mummies (Wyatt, 1970). The mummies may, however, be distributed with the cuttings so that the control, though inefficient, is self-perpetuating. Two parasites of *Macrosiphum euporbiae* are often present on tomatoes: *Praon volucre* and *Aphelinus asychis.* Although they have been tested in small-scale experiments, it has not proved possible to exploit them for large-scale biological control.

4.6b **POPULATION INTERACTIONS** For any natural enemy to control the numbers of its host, it must be capable of increasing more rapidly than the host. As we have seen, aphids are highly adapted reproductive organisms, achieving very high increase rates mainly on account of their rapid development – about a week under greenhouse conditions. Parasites, on the other hand, have development times of at least 2 weeks. Consequently, their increase rates of 2 to 6 × a week seldom equal those of aphids, which vary from 4 to 12 × a week.

On this basis it would hardly seem profitable to pursue the commercial use of parasites. However, there are several ways in which the action of parasites can be enhanced. Normally, control of aphids is required when their population density is low and their numbers are increasing at the maximum rate. As a population becomes more dense, the increase rate decreases and, providing the plant does not collapse, numbers stabilize at a steady density. Parasites are then able to overtake and bring about control (Wyatt, 1970). In most circumstances, the aphids would, by then, have reached damaging proportions (Figure 31T). However, on certain cultivars, for instance the 'Princess Anne' sports of chrysanthemum, *Myzus persicae* populations stabilize at a density of only about 10/plant and can be controlled by *A. matricariae* without undue damage (Figure 31A). Yet other chrysanthemum cultivars are so resistant to aphids that the parasites can increase more rapidly at all times and bring about rapid control (Figure 31P). Thus an ineffective parasite can be

FIGURE 31 Progress of populations of *Myzus persicae* and its parasite *Aphidius matricariae* on three chrysanthemum cultivars: 'Tuneful' (T); 'Golden Princess Anne' (A); 'Portrait' (P). Aphid populations: (□) observed; (—) calculated; (· · ·) theoretical maximum. Mummy numbers: (o) observed; (– – –) calculated.

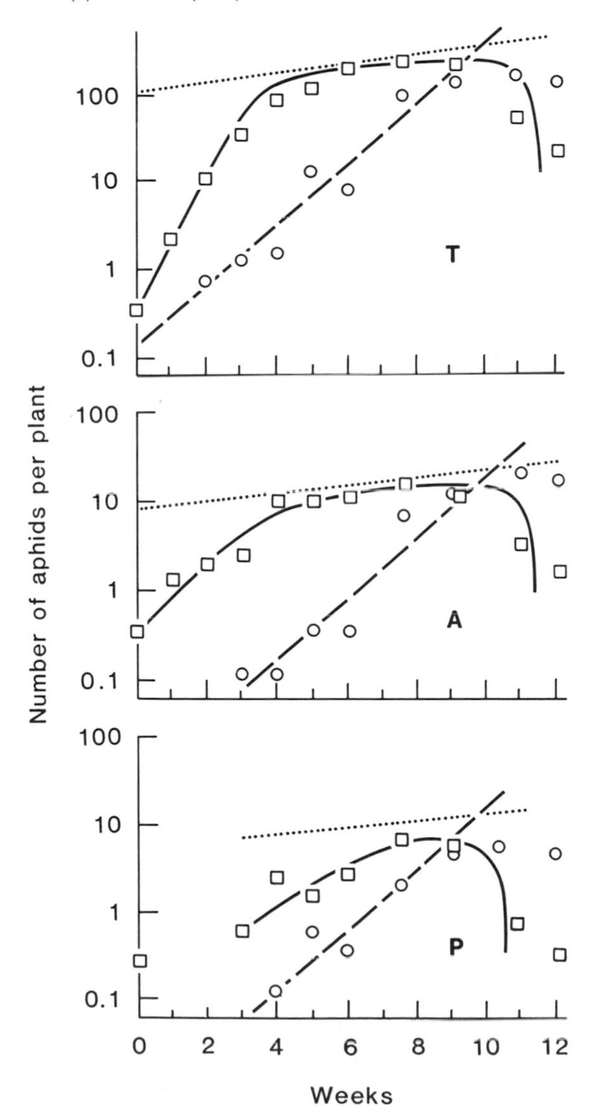

usefully used in conjunction with other methods of control, such as plant resistance or the use of predators (Lyon, 1979).

An alternative strategy is to introduce so many parasites that the increase rate of the aphids is greatly reduced from the start. *Aphis gossypii* can increase 12 × a week on cucumbers, but its most effective known parasite, *Aphelinus flavipes*, increases at only 6.3 × in the same conditions. It can be shown, in theory, that at least a tenth of the aphids must die each day from the outset to bring about control, and that about half the initial aphids must be parasitized to achieve this (Wyatt, 1983). Based on this concept, Scopes (1970) has suggested a method of controlling *M. persicae* on chrysanthemums by distributing highly parasitized populations of aphids in the boxes of cuttings.

4.6c **FURTHER BIOLOGICAL DISADVANTAGES OF PARASITES** Besides the more obvious disadvantages of slow increase rates, hymenopterous parasites suffer from several less obvious shortcomings. For instance, some species may not mummify the aphid until it has produced several young. It is only necessary for the aphids to produce more than one offspring each, on average, for the population to continue increasing whatever the parasite density (Scopes, Personal communication).

If parasites are introduced as mummies or adults to an established aphid outbreak, they cause no deaths until a week later and no further parasitism until 2 weeks. Thus there is an initial delay during which aphids can continue to increase at their maximum potential (Hofsvang & Hagvar, 1978 & 1980). Similarly, because parasites have a long development period and a relatively short period of egg-laying, their generations tend to be synchronized, allowing a lull of activity between generations when aphids can increase unchecked.

4.6d **CHOICE OF SPECIES** There are so many interacting considerations in the choice of potential aphid parasites that several approaches may be made:

1. A common species, known to thrive in greenhouse conditions may be chosen. *Aphidius matricariae* is a typical example attacking several aphid species.
2. Unusual or uncommon species may offer disadvantages. *Diaeretiella rapae*, a polyphagous parasite not often observed in greenhouses (Lyon, 1968), and *Ephedrus cerasicola*, an uncommon species (Hofsvang & Hagvar, 1978), have been tried against *M. persicae*.
3. Tropical species might be employed on the theory that they would be better adapted to the high temperatures and humidities of the greenhouse than the native aphids. *Aphidius colemani* has been introduced to Norway (Hofsvang & Hagvar, 1975) and *Lysiphlebus testaceipes*, which was introduced from Cuba to the south of France to control citrus aphids,

is now abundant there on populations of *Aphis gossypii* in greenhouses (Stary *et al.*, 1975).

4.6e **REARING PARASITES** Aphid parasites present no particular problems for mass-rearing. Being obligatory parasites, they must be raised on aphids, which, in turn, can only be reared on plants. Preferably they should be reared on the target aphid itself (Michel, 1972). Since efficient hyperparasites exist, rearing units should be small or ventilated with filtered air and carefully isolated.

The possibility of cold storage has been investigated to extend the production period. The mummies of *Ephedrus cerasicola* Hofsvang & Hagvar, 1978) and of *Aphidius matricariae* (Scopes *et al.*, 1973; Shalaby & Rabasse, 1979) can be cold-stored for at least a month.

REFERENCES

Hofsvang, T. & Hagvar, E.B. (1975) Duration of development and longevity in *Aphidius ervi* and *Aphidius platensis* (Hym.: *Aphidiidae*), two parasites of *Myzus persicae* (Hom.: *Aphididae*). *Entomophaga* **20**: 11–22.

Hofsvang, T. & Hagvar, E.B. (1978) Effect of parasitism of *Ephedrus cerasicola* Stary on *Myzus persicae* (Sulzer) in small glasshouses. *Z. Angew. Ent.* **84**: 1–15.

Hofsvang, T. & Hagvar, E.B. (1980) Use of mummies of *Ephedrus cerasicola* Stary to control *Myzus persicae* (Sulzer) in small glasshouses. *Z. Angew. Ent.* **90**: 220–226.

Lyon, J.P. (1968) Notes on using parasitic Hymenoptera for control of greenhouse aphids. *Annls Epiphyties* **19**: 113–118.

Lyon, J.P. (1979) Lâchers experimentaux de Chrysopes et d'Hyménoptères parasites sur pucerons en serres d'aubergines. *Annls Zool. Ecol. Anim.* **11**: 51–65.

Michel, M.F. (1972) Contribution a l'étude bioécologique des Aphelinides aphidiphages (Hym. Chalcidoidea). Thèse Doctorat d'Etat, Université de Paris VI. 124 pp.

Scopes, N.E.A. (1970) Control of *Myzus persicae* on year-round chrysanthemums by introducing aphids parasitised by *Aphidius matricariae* into boxes of rooted cuttings. *Ann. Appl. Biol.* **66**: 323–327.

Scopes, N.E.A., Biggerstaff, S.B. & Goodall, D.E. (1973) Cool storage of some parasites used for pest control in glasshouses. *Pl. Pathol.* **22**: 189–193.

Shalaby, F.F. & Rabasse, J.M. (1979) Effect of conservation of the aphid parasite *Aphidius matricariae* Hal. (*Hymenoptera: Aphidiidae*) on adult longevity, mortality and emergence. *Ann. Agric. Sci. Moshtohor* **11**: 59–73.

Stary, P., Leclant, F. & Lyon, J.P. (1975) Aphidiides (Hym.) et Aphides (Hom.) de Corse. I. Les Aphidiides. *Anns Soc. Ent. Fr.* **11**: 745–762.

Wyatt, I.J. (1970) The distribution of *Myzus persicae* (Sulz.) on year-round chrysanthemums. II. Winter season: the effect of parasitism by *Aphidius matricariae* Hal. *Ann. Appl. Biol.* **65**: 31–41.

Wyatt, I.J. (1983) Simple calculator models of predator–prey interactions: exponential population growth. *Protection Ecol.* **5**: 235–244.

4.7 APHID CONTROL BY FUNGI
R.A. Hall

4.7a ***VERTICILLIUM LECANII* (VERTALEC®) ON CHRYSANTHEMUM APHIDS** Following initial laboratory infectivity experiments and the development of methods of mass-producing infectious spores, experiments were undertaken to investigate the susceptibilities of various aphid pests on greenhouse crops. Aphids are most problematical on chrysanthemums; this crop is a particularly suitable subject for a fungal insecticide since, during the time of year when pests are most severe (April–September), the crop is covered at night with polythene blackouts (in temperate latitudes only) to restrict daylength and stimulate flowering. Under the blackouts, humidity rises to levels sufficiently high to permit fungal spore germination and penetration of the insect cuticle. In experimental greenhouse trials, it became apparent that *V. lecanii* possessed a remarkable capacity to control certain aphids (Hall & Burges, 1979; Hall, 1980). In descending order of susceptibility in greenhouse conditions, these species are: green-peach aphid (*Myzus persicae*), leaf-cutting plum aphid (*Brachycaudus helichrysi*), cotton aphid (*Aphis gossypii*), and black bean aphid (*A. fabae*). However, control has been variable with an occasional pest, the brown chrysanthemum aphid (*Macrosiphoniella sanborni*), which must sometimes be suppressed chemically (Hall & Burges, 1979). This is preferably done by spot-spraying with a selective aphicide like pirimicarb or, if need be, less specific chemicals such as nicotine, diazinon or dichlorvos can be used.

In larger-scale trials on commercial crops, consistently successful control of aphid species (other than *M. sanborni*) resulted in commercialization of the product in 1981 under the trade-name Vertalec® (Hall & Papierok, 1982). This product is manufactured by Tate & Lyle Ltd, Reading, England. Vertalec® is simply a dry white powder containing the same type of spores as are shown in their natural state in Figure 8, page 35.

A knowledge of the biology of the fungus helps us use the product in the best possible way. Firstly, dried spores of *V. lecanii* take longer to germinate than those which have been thoroughly wetted. Since spores can germinate and penetrate the cuticle only when humidity is high, usually during the limited period overnight, it is advantageous to soak the product in water for 2–3 hours prior to spraying, thereby effectively speeding up germination and hence the infection process. Secondly, *V. lecanii* requires about 3 weeks after spraying to control effectively an aphid population on chrysanthemums and eventually shrouds dead diseased aphids in a white, sporulating mycelium which adheres the bodies firmly to foliage. White aphid bodies on prominent foliage of an ornamental plant are aesthetically undesirable though they can be removed by oil or water sprays. To avoid this potential cosmetic problem, it is necessary to spray chrysanthemum plants early in the crop. So, if spores are sprayed 2–3 weeks after planting out cuttings, even when many aphids are present at the time of spraying, the marketable portion of the plant will have grown away from foliage bearing white bodies by the time the crop matures. Normally, however, aphid populations at the time of spraying are low. Some growers apply two sprays – one on the propagating bench and the second at the usual time, i.e. 2 weeks after planting. Two sprays maximize successful establishment of *V. lecanii* in the aphid population.

During 1984, commercial experience has shown that dry conditions can be counteracted by applying Vertalec® several times daily at rates from one tenth to one hundredth of the recommended dose. This method is particularly effective against *A. gossypii.*

Further recommendations may be made. Spray tanks should be rinsed out before applying Vertalec® so that any chemicals toxic to the fungus are removed. The fungus is best applied in the late afternoon, preferably just before the blackouts are pulled over, to minimise the possibilities of spores desiccating or of being injured by harmful ultra-violet rays.

The question of compatibility of Vertalec® with chemical insecticides and fungicides must also be considered. It is difficult to devise rapid screening methods whereby compatibilities of chemicals with the fungus can be realistically evaluated. To date, the effects of chemicals have been tested on fungal spore germination and growth *in vitro*, i.e. on agar and some chemicals, including some fungicides, were apparently safe. However, more recent and detailed studies have revealed that some of these 'safe' compounds may in fact reduce the effectiveness of the fungus when sprayed simultaneously. Therefore, a new recommendation has been proposed that, until further information is available, simultaneous applications of Vertalec® (or the whitefly strain, Mycotal®) with any chemical should be avoided. Certain fungicides, e.g. the dithiocarbamates, are very harmful and if possible should be avoided entirely. If they must be used such compounds should be applied at least 7 days or more *before* a Vertalec® application. A list of chemicals and their known effects on *V. lecanii* is given in Section 5.1, Table 4.

During the first year of commercial sales, results with Vertalec® were mixed, but following formulation changes (Hall & Turner, Unpublished), most results in 1982 and 1983 have been very satisfactory. Control of *Myzus persicae* is usually outstanding and presumably also that of *Brachycaudus helichrysi*, which is difficult to distinguish from *M. persicae.* Control of black aphid (*Aphis gossypii* and *A. fabae*) is generally good although a few colonies survive, clustered on individual plants, or on small groups, and should be destroyed with spot sprays of a chemical pesticide. Detection of these colonies demands only minimal vigilance on the part of the grower. The shiny brown chrysanthemum aphid (*Macrosiphoniella sanborni*), an occasional pest, is not very susceptible to Vertalec® in the greenhouse and should be controlled by pirimicarb or nicotine.

Possibly, in some parts of certain greenhouses, control may not be satisfactory; in one nursery, control adjacent to a south-facing concrete path (where presumably microclimate humidity was low) was poor but in the rest of the house was good. Such areas should be treated chemically.

Use of Vertalec® on chrysanthemums, strictly speaking, is recommended only during spring and summer months (April–September), when the use of polythene blackouts to restrict daylength has the effect of raising humidity at the same time. During the rest of the year, the performance of *V. lecanii* is less predictable; quite simply, the fungus seems to work in some growers' greenhouses and not others for reasons which are not as yet clear.

4.7b ***V. LECANII* (VERTALEC®) AGAINST APHIDS ON GREENHOUSE CROPS OTHER THAN CHRYSANTHEMUMS** On cucumbers, the major aphid pest is *A. gossypii*. The biotype on this crop is not yet known to be resistant to the selective aphicide, pirimicarb. However, since it is unwise to rely on a single pesticide to preserve the fragile balance of a complex integrated programme of control, the use of *V. lecanii* as an alternative, biological, strategy was evaluated. In experiments at the GCRI, control of this species on cucumbers could be obtained only by repeated application of spore sprays, in contrast to control of the same species on chrysanthemums where a single spore spray will give good control. The reasons for this difference are not clear but may be related to differences in temperature, microclimate humidity, behavioural differences between biotypes etc. in the two crops. Nevertheless, addition of a growth/germination substrate to the formulation encourages growth and sporulation on the leaf surface (invisible to the naked eye). Since spores not growing thus or not contacting an insect and growing on it soon dry and die, the effective inoculum potential may be increased by an order of 100 ×. A further advantage is that the spores are formed in protective mucilaginous heads and persist in a viable state for many weeks whereas sprayed spores become dried and fixed to the leaf surface and quickly die and therefore the effective increase in inoculum potential is probably many times more than 100. Using a spore-spray containing a substrate, control of *A. gossypii* on cucumbers can be obtained following just one spray.

On tomatoes, aphids are not a serious pest in Europe. However, in the USA, *Macrosiphum euphorbiae* can be troublesome. Experiments by R.

Lindquist, Ohio, showed that Vertalec® will control this species on this crop.

Vertalec® is compatible with the use of the braconid parasitoid (*Aphidius matricariae*), the red spider mite predator and parasitoids of leaf-miners.

To conclude, Vertalec® has a role to play in integrated control, especially on chrysanthemums; all the major aphid species are eradicated or suppressed and most growers have been satisfied with its performance. It is unlikely that Vertalec® will be used successfully in the extensive protected cropping areas of Southern Spain, Sicily, Crete and Colombia where the 'open' greenhouse structures create dry conditions and often the night-time temperature is low (in Colombia 0°–12°C depending on the time of year), factors which prevent successful establishment of the fungus. Few other fungi have been tested against aphids. In France, the entomophthoraceous fungi, *Conidiobolus obscurus* and *Erynia neoaphidis*, have been tested under glass (Dedryver & Rabasse, 1982) but generally results have been poor. However, in Colombia, in the savannah of Bogota, an entomophthoraceous fungus, probably *Zoophthora erinacea*, naturally controls aphids, albeit at high densities, on chrysanthemums (R.A. Hall, Personal observations). Doubtless, some attempt to augment levels of this fungus, earlier in the crop, would result in much better control.

REFERENCES

Dedryver, C.A. & Rabasse, J.M. (1982) Attempt at biological control of lettuce aphids in glasshouses with resting spores of *Conidiobolus obscurus* Hall and Dunn and mycelium of *Erynia neoaphidis* Remaud. and Henn. p. 103 in *Programme and Abstracts of Offered Papers, IIIrd International Colloquium on Invertebrate Pathology*, September 1982, University of Sussex, Brighton, UK.

Hall, R.A. (1980) Control of aphids by the fungus, *Verticillium lecanii*: effect of spore concentrations. *Entomologia Exp. Appl.* **27**: 1–5.

Hall, R.A. (1982) Control of whitefly, *Trialeurodes vaporariorum* and the cotton aphid, *Aphis gossypii*, in glasshouses by *Verticillium lecanii*. *Ann. Appl. Biol.* **101**: 1–11.

Hall, R.A. & Burges, H.D. (1979) Control of aphids in glasshouses with the fungus, *Verticillium lecanii*. *Ann. Appl. Biol.* **93**: 235–246.

Hall, R.A. & Papierok, B. (1982) Fungi as biological control agents of arthropods of agricultural and medical importance. *Parasitology* **84**: 205–240.

4.8 EXPERIENCE WITH THE SELECTIVE CONTROL OF CATERPILLARS USING *BACILLUS THURINGIENSIS*

P. Jarrett

The advantages of using the microbial insecticide, *Bacillus thuringiensis*, are many. It is specific in its action to caterpillars. It is harmless to Man, will not affect useful biological agents and is, therefore, suitable for integration into biological control systems. Its dosage levels are not limited and a crop can be harvested immediately after treatment. Commercial preparations of *B. thuringiensis*, of which several are available, e.g. Bactospeine®, Biobit®, Dipel® and Thuricide® contain, as their major active ingredients, crystals of toxic protein and bacterial spores.

On ingestion of toxic protein, caterpillars suffer paralysis of the mouthparts and gut muscles, resulting in a cessation of feeding. Death may occur in less than 48 hours. If too few toxin crystals are eaten to kill rapidly, the crystals still paralyse and disturb the gut so that spores can germinate, grow and cause death by septicaemia. Caterpillars not immediately eating a high dose may take up to 7 days to die, though feeding little in the meantime.

The main disadvantage of *B. thuringiensis* is that it has to be eaten by the caterpillar before it can take effect, there is no contact action. Consequently, thorough coverage of the crop is essential to obtain good control.

Different caterpillar species have different susceptibilities to *B. thuringiensis* so it is essential to identify the species involved. For this reason, the methods of application and the recommended application rates of the bacteria differ for individual crops.

4.8a **TOMATOES** The major caterpillar pest on tomatoes is the tomato moth (*Lacanobia oleracea*) and is the most susceptible of the glasshouse species to *B. thuringiensis*. HV sprays to run off at 0.1% w/v of *Bacillus* wettable powder give excellent control of caterpillar infestations (Jarrett & Burges, 1982). If an infestation comprises isolated patches on a few plants, spot spraying is adequate as eggs are laid in large batches, not deposited as single eggs throughout the crop.

As an alternative to applying the bacteria in HV sprays, which is both laborious and time-consuming, *B. thuringiensis* can be applied much more rapidly using pulse-jet-type fogging machines. The high temperatures generated by the machines do not harm the bacteria as the exposure period of the fog droplets to the high temperature is very short. Tests have shown that three pulse-jet-type machines, Pulsfog®, Swingfog® and Dynafog®, are suitable for use with the bacteria. Fogs of *B. thuringiensis* applied at the rate of 2.0 kg/ha for small and 4.0 kg/ha for large caterpillars give excellent control of the tomato moth (Burges & Jarrett, 1979) – providing adequate care is taken to obtain an even distribution of the fog throughout the crop. Water can be used as the fog carrier at a rate of 90 l/ha. Water avoids the expense of chemical carriers and the risk of phytotoxicity, although water should not be used as a carrier for chemical pesticides.

Measurements of thermal fog deposits of *B. thuringiensis* on leaf surfaces have shown that, on horizontal leaves, droplets fell only on the upper surface. The absence of deposit on the undersides of leaves is not a problem for the control of caterpillars, as they eat through to the upper surface, so obtaining a lethal dose of the bacteria. Applications of *B. thuringiensis* are active for more than 28 days, providing plants are not damped down or sprayed at high volume with other pesticides. Re-treatment is often unnecessary as caterpillars rarely move up the plants to feed on newly grown untreated foliage.

4.8b **CHRYSANTHEMUMS, PEPPERS AND LETTUCE** The three common caterpillar pest species found on these greenhouse crops are the angleshades moth (*Phlogophora meticulosa*), the silver-y moth (*Autographa gamma*) and the carnation tortrix moth (*Cacoecimorpha pronubana*). These three species are less susceptible to *B. thuringiensis* than the tomato moth so the dosage of bacteria has to be increased to 0.3% w/v to give good control. The bacteria should be applied as soon as caterpillars are found because early instars are more susceptible than later ones.

More than one application of bacteria is often necessary as caterpillars frequently move up the plants and feed on new untreated foliage. This behaviour is particularly prevalent on chrysanthemums where caterpillars move up the plant and feed on buds and flowers. Also, caterpillars of the carnation tortrix moth are often protected by leaves which they have folded together with silk. For this reason, it is advisable to repeat sprays until the pest cannot be found, as survivors can rapidly build up into a resident population within the glasshouse (Burges & Jarrett, 1976) and cause a more serious infestation than the original.

Thermal fogs of the bacteria are not recommended for these crops as droplets do not penetrate to the centre of chrysanthemum and pepper beds (Jarrett *et al.* 1978). Droplets are deposited only on the outer leaves of the crops. The only exception to the use of fogs on these crops is where plants are small and the dense foliage does not prevent penetration of droplets.

The cabbage moth (*Mamestra brassicae*), the large yellow underwing (*Noctua pronuba*) and the turnip moth (*Agrotis segetum*) are occasional pests of greenhouse crops. First instar caterpillars of the cabbage moth can be controlled by 0.6% w/v sprays of *B. thuringiensis*, but later instars and caterpillars of the other two species are relatively insensitive to and uncontrollable by the current commercial products of *B. thuringiensis*. If infestations need control, chemical insecticides such as diflubenzuron have to be used.

REFERENCES

Burges, H.D. & Jarrett, P. (1976) Adult behaviour and oviposition of five noctuid and tortricid moth pests and their control in glasshouses. *Bull. Ent. Res.* **66**: 507–510.

Burges, H.D. & Jarrett, P. (1979) Application and distribution of *Bacillus thuringiensis* for control of tomato moth in glasshouses. In *Proceedings of the 1979 British Crop Protection Conference – Pests and Diseases* **3**: 433–439.

Jarrett, P. & Burges, H.D. (1982) Control of tomato moth *Lacanobia oleracea* by *Bacillus thuringiensis* on glasshouse tomatoes and the influence of larval behaviour. *Entomologia Exp. Appl.* **31**: 239–244.

Jarrett, P., Burges, H.D. & Matthews, G.A. (1978) Penetration of controlled drop sprays of *Bacillus thuringiensis* into chrysanthemum beds compared with high volume spray and thermal fog. *Proceedings of the Symposium on Controlled Drop Application* 12–13 April 1979. Reading *Br. Crop Protect. Council Monogr.* No. 22: 75–81.

5. INTEGRATION OF BIOLOGICAL AND CHEMICAL CONTROL OF DISEASES AND MINOR PESTS

This section includes two contributions which to some extent overlap. They should be read together as they are complementary and contain information on different natural enemies.

5.1 TESTING THE SIDE EFFECTS OF PESTICIDES ON BENEFICIAL ORGANISMS BY OILB WORKING PARTY

S.A. Hassan & P.A. Oomen

When chemical pesticides are used to control pests and diseases, care should be taken to ensure that the most important beneficial organisms are not seriously affected. When beneficials are allowed to operate undisturbed in the field or in the greenhouse, less pesticide treatments may be needed to control pests. Studying the side effects of pesticides on natural enemies is particularly important when beneficials are introduced as biological control agents.

Evaluation of the side effects of pesticides on beneficial arthropods has attracted increasing attention by scientists in many parts of the world. Breeding techniques and standard test methods for important beneficials are being developed. The Working Group 'Pesticides and Beneficial Arthropods' of the International Organization for Biological Control (IOBC), West Palaearctic Regional Section (WPRS) was established in 1974 to encourage international co-operation. Beside the development of breeding and testing methods, the Working Group also organizes testing programmes in which pesticides of general interest are tested on a variety of beneficials and the results are published jointly (Franz *et al.*, 1980; Hassan *et al.*, 1983).

The present contribution summarizes techniques for testing the side effects of pesticides on beneficial organisms and includes the results of joint programmes carried out by the Working Group, with emphasis on greenhouse crops.

5.1a **TEST METHODS AND PROCEDURES** For practical reasons, there is general agreement among members of the Working Group that pesticides should first be tested in the laboratory on several beneficials relevant to the crops on which the pesticide is to be used before field tests are recommended. Pesticides found to be harmless to a particular beneficial in the laboratory are almost certain to be harmless to the same organism in the field and no further testing is done. Exceptions to this rule are few and do not justify a different approach. When a pesticide is found to be harmful in the laboratory, further testing in semi-field and field experiments is recommended.

5.1b **CHOICE OF BENEFICIALS FOR THE TEST** There is general agreement that the beneficials chosen should be relevant to the crops on which the pesticide is to be used. The alternative would be to test all pesticides on as many beneficials as is practical or on a fixed number of selected beneficials. The choice of beneficials is made with the aid of the following key:

1. Crops with aphids as pests: at least one aphid-predator (Chrysopidae, Coccinellidae, Syrphidae) as well as at least one aphid-parasite (Aphidiidae).
2. Crops with Lepidoptera as pests: at least one egg-parasite (Trichogrammatidae) as well as one larval-(pupal-) parasite (Tachinidae, Braconidae, Ichneumonidae).
3. Crops with mites as pests: at least one predator (*Typhlodromus* sp., *Amblyseius* sp. and/or Anthocoridae, for greenhouse crops *Phytoseiulus* sp.).
4. Crops with leaf-hoppers as pests: Anthocoridae.
5. Crops with flies as pests: at least one parasite (Cynipidae, Ichneumonidae).
6. When soil is treated: at least one predator (Carabidae, Staphylinidae) as well as one soil-living parasite (Cynipidae, Ichneumonidae).
7. Greenhouse crops with whiteflies as pests: the parasite *Encarsia formosa.*

Some examples of the choice of beneficials for testing FIELD CROPS Wheat: one general predator (Chrysopidae, Coccinellidae) one aphid-parasite (Aphidiidae), one soil-living predator (Carabidae, Staphylinidae).

VEGETABLES Brassica crops: one aphid-predator, one aphid-parasite, one Lepidoptera egg-parasite (*Trichogramma*), one Lepidoptera larval-(pupal-) parasite (Tachinidae, Braconidae), one soil-living predator, one soil-living parasite (Cynipidae, Ichneumonidae).

VEGETABLES UNDER GLASS Cucumber: the predatory mite *Phytoseiulus persimilis*, the whitefly parasite *Encarsia formosa*, one leaf-miner fly parasite, one aphid-parasite.

ORCHARDS Apple: one general predator, one aphid-parasite, one predatory mite (*Typhlodromus*), the mite-predator *Anthocoris*, one Lepidoptera egg-parasite, one Lepidoptera larval-(pupal-) parasite.
VINEYARDS The predatory mite *Typhlodromus pyri*, one Lepidoptera egg-parasite, one general predator, one predatory mite.

5.1c **STANDARD CHARACTERISTICS FOR TEST METHODS** Despite the diversity of the biology and behaviour of beneficial arthropods, members of the Working Group were able to agree on the following common features that characterize all tests: (a) laboratory, initial toxicity; (b) semi-field, initial toxicity; (c) semi-field, persistence. Characteristics for field test methods are still being discussed at the present time.

Laboratory, initial toxicity The procedures involved include: (a) exposure to fresh, dry pesticide films; (b) use of recommended concentration of pesticide; (c) application on glass plates, plants or soil (sand); (d) even film of pesticide, 1 to 2 mg fluid/cm^2; (e) laboratory reared arthropods, uniform age; (f) adequate ventilation; (g) water-treated controls; (h) four evaluation categories; (i) reduction in beneficial capacity/mortality.

Beneficials of uniform age are exposed to a fresh dry pesticide film applied at recommended concentrations on glass plates, plant leaves or soil, depending on the behaviour of the beneficials. The tests are carried out under controlled temperature and humidity conditions, favourable to the beneficial. Mortality in water-treated control units should not exceed certain limits. When closed cells are used, forced ventilation is provided to avoid the accumulation of pesticide fumes. Used air is pumped out of the cages so as to exchange the air every 2–3 minutes.

The pesticide solution, at recommended concentration, is sprayed to provide an even and reproducible film on the experimental surface. The amount of pesticide solution received on the experimental surface is measured by weighing the target before and immediately after spraying. At the end of an adequate exposure period, the evaluation is carried out by measuring the reduction in beneficial capacity, i.e. parasitism or egg-laying compared to controls. The pesticides are classified into four categories depending on the degree of damage they cause to the beneficial: 1 = harmless (< 50%), 2 = slightly harmful (50–79%), 3 = moderately harmful (80–99%) and 4 = harmful (> 99%).

Semi-field, initial toxicity Requirements: (a) exposure to fresh, dry pesticide films; (b) recommended concentrations of pesticide; (c) wet-spraying of plants; (d) field cages under field or field-simulated conditions; (e) water-treated controls; (f) laboratory-reared arthropods, uniform in age; (g) adequate contact through dense foliage; (h) food + host/prey near the centre of treated foliage; (i) adequate exposure period before evaluation; (j) four evaluation categories.

In this type of test, the laboratory-reared beneficials are exposed to a fresh, dry pesticide film applied on natural material (i.e. plant or soil) and the test units placed in the field under rain cover and partial shade or under field-simulated environment conditions. Plants should have a dense foliage to provide sufficient leaf surface for the pesticide treatment. Cages should be designed to suit the size and shape of the test plants. The plants, but not the experimental cages, are sprayed with the pesticide at recommended concentration to the point of run-off. After the pesticide has dried, the plants are caged, the beneficials are released and food is provided. The test units are placed in the field under transparent rain cover. Excessive heat and sunlight are excluded by providing partial shading. After an adequate exposure period, host or prey animals are placed among the treated plants and the performance of the beneficials is compared with control units treated with water.

Semi-field, persistence Requirements: (a) exposure to pesticide residues; (b) recommended concentrations of pesticide, (c) wet-spraying of plants; (d) weathering under field or field-simulated environmental conditions; (e) experiments up to 1 month after treatment; (f) laboratory-reared arthropods, uniform in age; (g) water-treated controls; (h) four evaluation categories.

The technique used to test the persistence (duration of harmful activity) of pesticide residues involves the treatment of plants or soil, maintaining them under field or field-simulated environment conditions, and exposing the beneficials to the treated substratum at different time-intervals after application. The 'duration of harmful activity' is the time required for the pesticide residue to lose effectiveness so that a reduction in parasitism of less than 50% compared with the control is reached. The pesticides are classified as follows: < 5 days duration of harmful activity: short lived; 5–14 days: slightly persistent; 15–30 days: moderately persistent; and > 30 days: persistent.

Field or greenhouse test Requirements: (a) crops inhabited by beneficials are directly treated; (b) laboratory-reared or naturally-occurring arthropods; (c) sampling at intervals before and after treatment(s); (d) recommended dose and number of treatments; (e) water-treated controls; (f) dead and/or living individuals are collected; (g) number of individuals must exceed a certain limit to allow statistical analysis.

Beneficials that occur in sufficient numbers in the field, or natural enemies that are mass-reared in insectaries, i.e. for use in biological control, can easily be tested under practical conditions. These tests involve the treatment of crops inhabited by beneficials with the pesticide to be treated at recommended doses and monitoring the effects. Samples of dead and/or living individuals are taken at intervals before and after treatment(s). Deviations from water-treated control plots are used to measure the effect of the chemical on beneficial species of group of species.

5.1d **RESULTS** Results of experiments to test the side-effects of 40 pesticides on 9 to 13 beneficial arthropods carried out by members of the Working Group in 6 different countries using standard test methods were published by Franz *et al.* (1980) and Hassan *et al.* (1983). The beneficial arthropods tested were *Amblyseius potentillae*, *Anthocoris nemorum*, *Chrysopa carnea*, *Coccygomimus turionellae*, *Cryptolaemus montrouzieri*, *Drino inconspicua*, *Encarsia formosa*, *Leptomastix dactylopii*, *Opius* sp., *Phygadeuon trichops*, *Phytoseiulus persimilis*, *Syrphus vitripennis*, *Trichogramma cacoeciae*. The insecticides Dipel®, Torque®, Dimilin®, Azomate®, the fungicides Nimrod®, Cerocbin-M, Ortho Difolatan®, Orthocid 83®, Bayleton®, Ronilan®, Derosal®, the herbicides Illoxan®, Kerb 50 W® and Semeron® were harmless to most of the beneficials tested and can be recommended for use in integrated control programmes. The remaining 16 insecticides, 5 fungicides and 5 herbicides were harmful or moderately harmful to most of the test organisms. Further testing of these preparations using standard semi-field or field test methods is recommended.

Table 4 (pp. 150–151) shows the side-effects of pesticides commonly used for pest control on glasshouse crops in Europe on the whitefly parasite (*Encarsia formosa*), the predatory mite (*Phytoseiulus persimilis*), the leaf-miner parasites (*Diglyphus* sp., *Dacnusa* sp., *Opius* sp.) and the entomophagous fungus (*Verticillium lecanii*). As an indication of the duration of harmful activity (persistence) results of a semi-field test with *Trichogramma cacoeciae* were included. Fungicides based on triforine, vinclozolin, iprodione, bupirimate were harmless to all the beneficial organisms tested; imazalil and zineb were harmless to *E. formosa* and *P. persimilis* and ought to be further tested.

REFERENCES

Franz, J.M., Bogenschutz, H., Hassan, S.A., Huang, P., Naton, E., Suter, H. & Viggiani, G. (1980) Results of a joint pesticide programme by the Working Group: Pesticides and Beneficial Arthropods. *Entomophaga* **25** (3): 231–236.

Hassan, S.A. (1982) Relative tolerance of three different strains of the predatory mite *Phytoseiulus persimilis* to 11 pesticides used on glasshouse crops. *Z. Angew. Ent.* **93**: 55–63.

Hassan, S.A. (1983) Ergebnisse der Laborprüfung einer Reihe von Pflanzenschutzmitteln auf Eiparasiten der Gattung *Trichogramma* (Hymenoptera, Trichogrammatidae). *Nachbl. Dt. Pflschutzdienst. Berl.* **35**: 21–25.

(cont'd p. 152)

TABLE 4 Side effects of pesticides on beneficial organisms that are used to control pests of glasshouse crops. The side effects are measured by initial toxicity texts in the laboratory.

Pesticide	Concentration (ppm active ingredient)	Side effect on beneficial organism in classes[1]						
		Encarsia formosa Adults	*Encarsia formosa* Pupae	*Encarsia formosa* Persistence[2]	*Phytoseiulus* Juv.–Adults	*Diglyphus* Adults	*Dacnusa/Opius* Adults	*Verticillium lecanii*
INSECTICIDE/ACARICIDE								
carbaryl	1062	4	3	11+	4	—	4	—
cyhexatin	250	4	1	31+	4	1	4	1
cypermethrin	50	4	3	21+	—	—	—	—
deltamethrin	12.5	4	4	—	4	—	—	1
diazinon	142	4	1	7	4	4	4	—
dicofol	278	3	1	13	2	1	—	—
dienochlor	500	3	1	—	4	—	—	1
endosulfan	1089	4	1	20	4	—	—	—
fenbutatin oxide	250	1	—	—	1	—	—	—
heptenophos	275	4	1	4	4	—	3	—
lindane	140	4	1	7	1	—	3	—
malathion	750	4	4	10+	4	—	—	—
methomyl	250	4	3	31+	3	—	4	—
mevinphos	72.5	4	1	3−	—	—	—	—
oxamyl	500	4	2	—	1	—	—	—
parathion	150	4	2	14	4	—	—	—
permethrin	37.5	4	2	31+	4	—	3	1
pirimicarb	250	4	1	1	3	1	3	1
pyrimiphos-methyl	500	4	2	4	4	—	4	—
proclonol	285	2	1	—	—	—	—	—
propoxur	375	4	4	11	3	—	3	—
pyrethrum/ piperonylbutoxide	37.5	4	3	9	—	—	—	—
trichlorphon	800	4	1	—	4	—	3	—

FUNGICIDE								
benomyl	200	1	—	—	4	—	—	4
bitertanol	300	1	—	—	—	—	—	—
bupirimate	100	1	—	—	1	—	—	1
carbendazim	360	1	—	—	3	—	1	—
chlorothalonil	1500	1	—	—	1	—	—	4
dichlofluanid	1000	3	1	22	1	—	—	4
dinocap	150	2	1	8	1	—	—	1
ditalimfos	375	3	1	19	1	—	1	—
fenarimol	24	1	—	—	2	—	—	1
imazalil	50	1	—	—	1	—	—	—
iprodione	500	1	—	—	1	1	1	1
maneb	1600	4	1	—	1	—	—	4
oxycarboxin	300	—	—	—	—	1	1	—
procymidon	132	1	—	—	—	—	—	—
pyrazofos	120	4	1	31+	1	—	—	1
thiophanate-methyl	350	1	—	—	2	—	—	—
thiram	1600	2	1	31+	1	—	—	4
tolylfluanid	1250	4	1	—	—	—	—	—
triforine	200	1	—	—	1	1	1	1
vinclozolin	250	1	—	—	1	1	1	1
zineb	1400	1	—	—	1	—	—	—
maneb/vinclozolin	960/150	4	1	—	—	—	—	—

[1]The information presented in this table was collected from Franz et al. (1980), Hassan (1982), Hassan (1983), Hassan et al. (1983), Ledieu & Helyer (1983), Oomen & Wiegers (1983), Overmeer & van Zon (1983), Samsøe-Petersen (1983). The side effect is expressed in standardized IOBC- classes:

Class	Evaluation	% Reduction in parasitization/predation
1	harmless	0–49
2	slightly harmful	50–79
3	moderately harmful	80–99
4	harmful	>99

[2]Persistence of the pesticide on leaves in semi-field tests (LT 50 = number of days until residue is sufficiently aged to kill at most half of all *Trichogramma cacoeciae* adults exposed. *Trichogramma* and *Encarsia* show the same pesticide sensitivity).

Hassan, S.A., Bigler, F., Bogenschutz, H., Brown, J.V., Firth, S.I., Huang, P., Ledieu, M.S., Naton, E., Oomen, P.A., Overmeer, W.P.J., Rieckmann, W., Samsøe-Petersen, L., Viggiani, G. & van Zon, A.Q. (1983). Results of the second joint pesticide testing programme by the IOBC/WPRS Working Group 'Pesticides and Beneficial Arthropods'. *Z. Angew. Ent.* **95**: 151–158.

Ledieu, M.S. & Helyer, N. (1983) *Guidelines for Integrating Pesticides with Natural Enemies.* Unpublished data, Glasshouse Crops Research Institute, Littlehampton, UK.

Oomen, P.A. & Wiegers, G.L. (1983) Unpublished data. Plant Protection Service, Wageningen.

Overmeer, W.P.J. & van Zon, A.Q. (1983) Unpublished data. Laboratory for Experimental Entomology, Amsterdam.

Samsøe-Petersen, L. (1983) Laboratory method for testing side-effects of pesticides on juvenile stages of the predatory mite, *Phytoseiulus persimilis* (Acarina, Phytoseiidae) based on detached bean leaves. *Entomophaga* **28** (2): 167–178.

5.2 EVALUATION OF SIDE-EFFECTS OF PESTICIDES BY THE GLASSHOUSE CROPS RESEARCH INSTITUTE

M. Ledieu

5.2a **METHODS OF TESTING** Screening of pesticides for adverse effects on beneficial organisms can be done at several different levels, according to requirements. Where information must be obtained quickly, a very simple test based on mortality of beneficials is adequate; alternatively, more complex tests may measure the effect of the pesticide on performance of the beneficial or on other parameters of interest.

The first approach was adopted at the GCRI because growers were already using beneficials in the absence of any information to guide them about which pesticides were suitable for integration and which were not. Thus, simple mortality tests were developed (Ledieu, 1979; Helyer, 1982; Ledieu & Helyer, 1983) for use in the laboratory and were supplemented, where necessary, by tests in the greenhouse on crop plants. The policy was to provide simple information quickly, which could be used by the growers as a guide in decision-making, and to rely on the growers to provide feed-back in the case of materials which were unexpectedly harmful. In this way, the more laborious greenhouse tests were mainly restricted to 'problem' pesticides.

1. **Simple laboratory test for parasites** The adult stage is likely to be the most sensitive to pesticides so this was the stage chosen for testing. A simple cage structure was made by using thick capillary matting material as a spacer between two sheets of glass (microscope slides). The matting was 7–8 mm thick and a circular hole was made through it (*c.* 20 mm in diameter) to form the test chamber. A treated leaf disc was placed over the lower piece of glass and the matting was then placed over the disc. About 10–15 parasites, collected in gelatin capsules, were then immobilized by placing them in the freezer compartment of a fridge for *c.* 1 minute. When they had been introduced into the test cage, together with a minute piece of filter paper soaked in 10% sucrose, the top piece of glass was put in place and the whole assembly held together with elastic bands. As soon as parasites recovered from immobilization, they were examined so that mortality due to damage during handling could be subtracted from subsequent mortality due to pesticide action. Mortality assessments were made twice per day until mortality in control cages exceeded 10%.

Leaf discs were sprayed using a Potter tower operating at 40 kPa pressure. 2 ml of water (controls) or of pesticide (at either 10 × commercial recommended concentration (*Rc*), or *Rc*, or 0.1 × *Rc*) were sprayed onto leaf discs. Pesticides expected to be harmful were tested initially at $0.1 \times Rc$ and Rc, whereas those expected to be safe were tested at Rc and $10 \times Rc$.

Persistence of spray deposits can be assessed by spraying a plant at Rc and removing leaf discs for testing at various intervals thereafter.

Materials that were found to be harmful in the above test were then used in further tests in which their effect on the pupal stage was assessed. Leaf portions bearing about 150 pupae were dipped into pesticide solutions (concentrations as above). These were then kept at 20°C with a 16 hour photoperiod until adults emerged.

2. **Simple laboratory test for predators** In this test, leaf portions (*c.* 15–20 cm^2) bearing prey of all ages, and with webbing removed if too much is present, were placed on sodden cotton wool to prevent escape of prey or predator. Ten adult predators were placed on the leaf and it was then sprayed as above. Mortality of predators was assessed as above and counts of predator eggs laid were made at the same time. Persistence of spray residues was assessed as above.

Slight variations of the above test allow other aspects of a pesticide's activity against predators to be studied. The above method tests for combined contact, residue and food-chain activity but other stages can be used in place of adults or predators.

3. **Greenhouse tests** These are intended to simulate commercial conditions and involve setting up the pest and biological control agent on crop plants according to classical 'pest-in-first' techniques. Once a short period to allow establishment has elapsed the crop is then sprayed weekly at $2 \times Rc$ (the recommended dose) and the progress of pest and beneficial populations is compared with those on water-sprayed control plants.

5.2b **INTERPRETATION OF RESULTS** The broad aim of the simple tests set out above is to enable pesticide formulations to be allocated to one of three groups. The two extreme groups are 'harmful' and 'safe' and the tests are designed so that if a pesticide falls into either of these groups then there is no doubt as to the course of action to take. If more than 50% of test organisms are killed by $0.1 \times Rc$ the pesticide is classed as Harmful. Conversely, if less than 50% are killed by $10 \times Rc$ the material is Safe. Pesticides which do not fall into either of the above groups are classified as 'intermediate' and in this case the action to be taken is much less definite; it is for the grower to decide, on the basis of his own circumstances, whether the risk of harming his beneficials is acceptable or not. Once the decision has been taken to use beneficials, the process of selecting the most suitable pesticides for integration involves four steps (go to p. 160).

TABLE 5 Guidelines for use of pesticides with natural enemies*.

Pesticide	Applied as	*Phytoseiulus* Eggs	*Phytoseiulus* Adults	*Encarsia* Pupae	*Encarsia* Adults	*Diglyphus* Adults	*Dacnusa/ Opius* Adults	*Verticillium lecanii*
INSECTICIDES								
aldicarb	granules	H	H	—	—	—	—	—
Bacillus thuringiensis	HV spray	S	S	—	S	—	—	—
bioresmethrin	HV spray	S	I	S	H3	—	—	—
carbaryl	HV spray	—	S	H	H11 +	—	H	B
chlorpyrifos	HV spray	—	H	I	H	—	—	—
cypermethrin	HV spray	—	—	H	H21 +	—	—	—
DDT	HV spray	—	H	H	H	—	—	—
DDT	smoke	—	—	—	—	—	—	—
DDT/γ-HCH	smoke	—	—	H	H10	—	—	—
deltamethrin	HV spray	H	H	H	H	—	—	B
demeton-S-methyl	HV spray	H	H	H	H7	—	—	—
demeton-S-methyl	soil drench	—	H	—	H70	—	—	—
derris	HV spray	I	H	S	H1	—	—	—
dialifos	HV spray	—	—	—	H	—	—	—
diazinon	HV spray	—	H	H	H7	H	H	—
diazinon	soil drench	—	S	S	—	—	—	—
diazinon	ULV spray	—	—	—	—	—	—	—
dichlorvos	HV spray	—	H	H	H2	—	—	—
dichlorvos	ULV spray	—	—	—	—	—	—	—
diflubenzuron	HV spray	S	S	—	S	—	—	B
dimethoate	HV spray	H	H	H	H21	—	—	—
dimethoate	soil drench	H	—	—	—	—	—	—
dimethoate	ULV spray	—	—	—	H60	—	—	—
dimethoate	Turbair	H	H24	I	H14	—	—	—
dioxathion	HV spray	—	H	I	H	—	—	B

TABLE 5 Guidelines for use of pesticides with natural enemies*. (cont'd).

Pesticide	Applied as	*Phytoseiulus* Eggs	*Phytoseiulus* Adults	*Encarsia* Pupae	*Encarsia* Adults	*Diglyphus* Adults	*Dacnusa/Opius* Adults	*Verticillium lecanii*
endosulfan	HV spray	—	H4	—	I	—	—	—
fenpropathrin	HV spray	—	H	—	—	—	—	—
formothion	HV spray	I	H	—	—	—	—	—
γ-HCH	HV spray	H	H4−	H	I4−	—	—	—
γ-HCH	soil drench	—	I	S	—	—	—	—
γ-HCH	ULV spray	—	—	—	—	—	—	—
γ-HCH	smoke	S	H	—	—	—	—	—
heptenophos	HV spray	S	H3	S	I4	—	I	—
malathion	HV spray	H	H10	H	H10+	—	—	—
malathion	ULV spray	—	—	—	—	—	—	—
mecarbam	HV spray	—	H	—	H	—	—	—
methomyl	HV spray	—	H	—	—	—	—	—
monocrotophos	HV spray	—	H	I	H120	—	—	—
nicotine	HV spray	—	I4−	S	H4−	—	—	—
nicotine	smoke	S	S	S	H	—	—	—
oxamyl	granules	S	S	—	—	—	—	—
oxydemeton-methyl	HV spray	—	H	I	—	—	—	—
oxydemeton-methyl	soil drench	—	—	—	—	—	—	—
parathion	HV spray	H	H14	H14	H	—	—	—
parathion	soil drench	H	H14	H14	—	—	—	—
parathion	smoke	H	H14	H14	—	—	—	—
permethrin	HV spray	—	H	S	H	—	—	B
permethrin	Turbair	H	H28	H	H21	—	—	—
permethrin	fog	—	H	—	H15+	—	—	—
pirimicarb	HV spray	S	S	S	I	S	I	B
pirimicarb	smoke	—	—	—	—	—	—	—
pirimiphos-methyl	fog	—	H	H	H16	—	—	—

pirimiphos-methyl	smoke	—	—	—	—	—	—	—
propoxur	smoke	—	H8	S	—	—	—	—
pyrethrum	HV spray	—	—	H	H8	—	—	—
pyrethrum	fog	—	—	S	H1	—	—	—
resmethrin(Extra)	Turbair	H	H7	S	H3	—	—	—
resmethrin/pyrethrum	HV spray	H	H4 –	–	H4 –	—	—	—
resmethrin/pyrethrum	Turbair	H	H7	S	H3	—	—	—
resmethrin/pyrethrum	ULV spray	H	H	S	—	—	—	—
resmethrin/pyrethrum	fog	—	H	I	H	—	—	—
rotenone/pyrethrum	HV spray	—	I	—	H	—	—	—
rotenone	HV spray	I	H	S	H1	—	—	—
schradan	HV spray	—	H	—	—	—	—	—
sulfotep	smoke	—	H	—	—	—	—	—
TEPP	HV spray	—	H	—	—	—	—	—
trichlorphon	HV spray	—	H	I	H	—	—	—
FUNGICIDES								
benodanil	HV spray	S	S	—	S	—	—	B
benomyl	HV spray	H	H	—	S	—	—	A
benomyl	soil drench	I	I	S	S	—	—	B
bupirimate (EC)	HV spray	I	S	I	—	I	S	B
bupirimate(WP)	HV spray	S	S	—	S	—	—	—
captafol	HV spray	S	S	—	S	—	—	—
captan	HV spray	S	S	S	S	—	—	A
carbendazim	HV spray	S	I	S	S	—	—	—
chlorothalonil	HV spray	S	S	S	S	—	—	A
chlorquinox	HV spray	S	S	—	S	—	—	—
copper oxychloride	HV spray	S	S	S	S	—	—	—
copper oxychloride	Turbair	S	S	S	S	—	—	—
cupric ammonium carb.	HV spray	S	S	S	I	—	—	—
dichlofluanid	HV spray	—	S	S	S	S	—	A
dicyclidine	HV spray	S	S	—	S	—	—	—
dimethirimol	HV spray	—	S	S	S	—	—	—
dinocap	HV spray	—	S	S	H	—	—	B

TABLE 5 Guidelines for use of pesticides with natural enemies*. (cont'd).

Pesticide	Applied as	*Phytoseiulus* Eggs	*Phytoseiulus* Adults	*Encarsia* Pupae	*Encarsia* Adults	*Diglyphus* Adults	*Dacnusa/Opius* Adults	*Verticillium lecanii*
ditalimphos	HV spray	S	S	—	—	—	—	—
drazoxolon	HV spray	—	S	—	—	—	—	—
etridiazole	HV spray	—	—	—	S	—	—	—
fenarimol	HV spray	—	I	—	S	—	—	B
fentin acetate	HV spray	—	—	—	S	—	—	—
fluotrimazole	HV spray	S	S	—	S	—	—	—
imazalil	HV spray	I	S	H	I	—	—	A
iprodione	HV spray	S	S	I	S	S	S	B
iprodione	Turbair	S	S	S	S	—	—	—
mancozeb/zineb	HV spray	—	S	—	—	—	—	—
maneb	HV spray	—	S	S	I	—	—	A
metalaxyl	HV spray	—	H	—	S	—	—	—
nabam	HV spray	—	—	S	I	—	—	—
nitrothal-isopropyl/Zn	HV spray	—	—	—	S	—	—	—
nitrothal-isopropyl/S	HV spray	—	I	—	S	—	—	—
oxycarboxin	HV spray	S	S	S	S	S	S	B
prochloraz(EC)	HV spray	—	H	—	—	—	—	—
prochloraz(WP)	HV spray	—	S	—	—	—	—	—
pyracarbolid	HV spray	—	—	—	S	—	—	—
pyrazophos	HV spray	H	I4	H	H	—	—	B
quintozene	HV spray	—	—	—	S	—	—	—
thiophanate-methyl	HV spray	H	S	—	S	—	—	—
thiram	HV spray	I	S	—	S	—	—	A
triadimefon	HV spray	S	S	—	S	—	—	—
triforine	HV spray	S	S	I	S	S	S	B
vinclozolin	HV spray	—	S	S	S	S	S	B
zineb	HV spray	—	S	—	S	—	—	—
zineb	Turbair	S	S	S	S	—	—	—

ACARICIDES								
amitraz	HV spray	H	H	H	H	—	—	—
cyhexatin	HV spray	I	S	S	S	S	H	B
dicofol (emulsion)	HV spray	—	H	S	S	—	—	—
dicofol(EC)	HV spray	H	H	—	—	S	—	—
dicofol	Turbair	H2	S	S	S	—	—	—
dicofol/tetradifon	HV spray	I	I	—	I	—	—	—
dicofol/tetradifon	Turbair	H2	S	S	S	—	—	—
dicofol/tetradifon	fog	—	I	—	S	—	—	—
dienochlor	HV spray	I	H	S	S	—	—	B
dinobuton	HV spray	S	S	—	—	—	—	—
fenbutatin oxide	HV spray	—	S	—	S	—	—	—
fentrifanil	HV spray	—	H	S	I	—	—	—
petroleum emulsion	HV spray	S	I	I	I	—	—	B
propargite	HV spray	H	—	I	I	—	—	—
propargite	Turbair	H	H	I	I	—	—	—
quinomethionate	HV spray	H	H	S	I	—	—	—
quinomethionate	smoke	H	H	—	—	—	—	—
tetradifon	HV spray	S	S	S	S	S	S	B

*These data are based on laboratory tests and may, therefore, not be a true reflection of what might happen in commercial greenhouses. They are intended to be used as a *guideline* in the absence of practical experience:

H = harmful
I = intermediate
S = safe
H70 = harmful for 70 days
H5+ = harmful for more than 5 days
H2− = harmful for less than 2 days
A = incompatible: apply at least 7 days after *V. lecanii* but preferably at least 7 days before
B = compatible except when tank-mixed

1. Decide which beneficials will be needed. It is very important to do this at the outset, otherwise a decision to include another beneficial at a later date might be impractical because pesticides already in use are compatible with existing beneficials but lethal to others.
2. Make a comprehensive list of effective pesticides for each potential pest.
3. Beside each chemical in the list write in the effects on your chosen beneficials as listed in Table 5. Where a 'dash' appears in the Table the pesticide has not been tested. Many of these gaps can be filled by entering 'safe' to *Encarsia* pupae where the pesticide is 'safe' to adult *Encarsia.* This is permissible because the pupae are protected by their waxy case. For remaining gaps, play safe and assume that the pesticide is 'harmful', unless there is good evidence to the contrary.
4. It should now be possible to select the most innocuous pesticide for integrated control of each pest. If not, then the information on persistence of harmful effects, given in Table 4 (p. 155), should be considered and the least persistent material selected. Where no persistence information is given, substitute the 'minimum harvest interval' listed in the British Agricultural Chemical Approval Scheme Handbook *Approved Products for Farmers and Growers.*

There are a number of other factors that must be considered before a pesticide is used:
1. Damage will be greater when pesticides are applied before thorough establishment of beneficials. Try to ensure that only 'safe' pesticides are used during the early stages of an interaction.
2. Frequent sprays are bound to be more harmful than occasional ones.
3. The application method (e.g. HV spray or soil drench) can influence the effect of a pesticide on beneficials (see benomyl *v. Phytoseiulus* or demeton-S-methyl *v. Encarsia*).
4. The persistence of some pesticides is longer during dull winter weather than in hot sunny weather.
5. Where there is no alternative but to use a pesticide which is likely to be very damaging to beneficials, care must be taken to apply the material so as to separate the residue from the niche occupied by the beneficial.
6. Finally, remember that the information presented in Table 5 is derived from simple tests which take account only of mortality effects. It should, nevertheless, give a reasonable indication of the outcome of using a pesticide in the greenhouse.

REFERENCES

Anon. (1983) List of approved products and their uses for farmers and growers. Reference book **380** (3). HMSO, London. 246 pp.

Helyer, N.L. (1982) Laboratory screening of pesticides for use in integrated control programmes with *Encarsia formosa* (Hymenoptera, Aphelinidae) and *Phytoseiulus persimilis* (Acarina, Phytoseiidae). In *Tests of Agrochemicals and Cultivars Ann. Appl. Biol.* **100** (3): 64–65 (Supplement).

Ledieu, M.S. (1979) Laboratory and glasshouse screening of pesticides for adverse effects on the parasite *Encarsia formosa* Gahan. *Pestic. Sci.* **10**: 123–132.

Ledieu, M.S. & Helyer, N.L. (1983) Integration of pesticides with biological control agents. p. 1111. In *Proceedings of the Xth International Congress of Plant Protection* 20–25 November 1983, Brighton, UK.

5.3 THE ECOLOGICAL SELECTIVITY OF PESTICIDES IN INTEGRATED PEST MANAGEMENT
N.L. Helyer

Chemicals used in integrated control programmes must perform two functions: (a) control the pest or disease outbreak and (b) have only a limited effect on any parasites or predators being used. Any overall effects must favour the natural enemy rather than the pest. The selective use of chemicals with natural enemies is achieved in three ways: (a) true selectivity – chemicals are used for their safety to beneficial insects whilst being effective against the pest (pirimicarb against aphids); (b) separation in time, whereby chemical applications are timed so that harmful effects are negligible to the natural enemies whilst still being effective against the pest (when a parasite is protected within its host); (c) separation in space, whereby chemical treatments are separated from beneficial insects by their application methods (polybutene + insecticide sprayed below a crop to kill insects which migrate to the ground to pupate).

5.3a **TRUE SELECTIVITY** The aphicide, pirimicarb, has minimal effects on natural enemies and is safe to use on most horticultural crops; it is also available to the amateur gardener in the UK. It is usually applied as an HV spray but it can be also used as a soil drench to control aphids by its systemic activity.

Another example of a useful chemical is diflubenzuron which can be used as a soil drench for pot plants to control larvae of sciarid fly and also be applied as a foliar spray against many species of caterpillar. Diflubenzuron acts by interfering with the deposition of chitin, one of the main constituents of the insect cuticle. After contamination (ingestion), the malformed cuticle of the new instar cannot withstand the internal pressures during ecdysis so that the larvae is unable to cast its old skin.

The bacterial pathogens, such as *Bacillus thuringiensis*, and the fungal spores of *Verticillium lecanii* (p. 142 and p. 116) can be safely used in integrated control programmes as they attack only their target hosts.

Another example of selectivity is the selection of insects resistant to certain groups of chemicals. Pesticide pressure has been applied to pests for years and more recently, through integrated control programmes, to the natural enemies. At present the predatory mite, *Phytoseiulus persimilis*, has developed resistance to both organophosphorous and carbamate chemicals. It appears that the levels of resistance vary throughout Europe, apparently depending on the pesticide pressures that have been applied.

It is not certain how to maintain or develop resistance in *P. persimilis* during mass production but in Europe resistance has been found to diazinon, pyrazophos and carbaryl. This resistance is being exploited by commercial suppliers of natural enemies who are applying regular sprays of the appropriate chemical to try and maintain, or increase, this resistance in the natural enemies they supply.

There is a continuing programme throughout the world to improve the performance of *P. persimilis*; Voroshilov (1979) selected for high temperature tolerance while Schulten & van de Klashorst (1979) obtained high levels of resistance to parathion. Wearing and his group in New Zealand have started a programme of selection for pyrethroid resistance.

5.3b **SEPARATION IN TIME** The correct timing of pesticide applications can be very important in preserving the balance between pest and prey. Control of whitefly (*Trialeurodes vaporariorum*) by *Encarsia formosa* on early (October–November) sown tomatoes is impractical due to low greenhouse temperatures. Experimentally, the use of yellow stick traps at 1/4.5 m^2 of bench, together with oxamyl treatment of the young plants gives adequate control. The oxamyl treatment was carried out during the first week of December when the plants were stood out on the peat-modules. This permits planting in early January without risk and ensures that any whitefly subsequently invading the crop will not produce scales suitable for parasitism before the middle of February, when introductions of *E. formosa* can commence. Alternatively, oxamyl can be applied twice at 4-week intervals, making the last treatment 6–8 weeks before introductions so that no toxic honeydew is produced.

Similarly chemical control may be necessary for spider mites on young chrysanthemum plants until the leaf canopy is touching and thus enabling the predatory mite, *P. persimilis*, to search effectively for any invading spider mites. Natural enemies that fly are obviously not hindered in this way, but they are generally less efficient at finding their host during the winter due to low temperatures so again chemical control may be necessary.

The larval stages of most parasites are protected from pesticides within the host or 'mummy'; it is therefore useful to know when this stage occurs in order to time any pesticide applications. The use of the aphid parasite, *Aphidius matricariae*, to control *Myzus persicae* on chrysanthemums is one example of where this can be used. The rate of development of an insect within normal greenhouse conditions is almost linearly related to temperature. Scopes & Biggerstaff (1977) found that development of *A. matricariae* from egg to adult required 161.6 ± 2.1 day-degrees C with a threshold temperature (below which development all but ceases) of 7.9°C.

A temperature integrator is a digital counter whose counting is proportional to the temperature above a threshold in an analogous way to the rate of insect development. Modern integrators can be computer-linked to provide direct readings of mean temperatures, as well as maximum and minimum figures over a prescribed period, thus forecasting when a 'safe' period will occur.

Heptenophos is a systemic organophosphorous insecticide of short persistence used as an aphicide. It can also be used to control leaf-hopper and leaf-miner and gives some control of thrips. However, due to its short persistence, it can be integrated with *E. formosa* and *P. persimilis*. The larval stages of *E. formosa* are protected within the host scale but the adult is vulnerable, so, if an established population is sprayed the adults are killed, but after 2 to 3 days newly emerging adults remain unharmed. Similarly, with *P. persimilis*, the adults and mobile young are vulnerable but the eggs are safe and after 2–3 days can survive on treated plants.

5.3c **SEPARATION IN SPACE** Most tomato and cucumber crops are not grown directly on the soil but in an isolated medium standing on polythene sheeting, e.g. growbags, rock-wool and nutrient film. These new growing techniques eliminate the need to sterilize the soil but they also encourage populations of soil-pupating insects to overwinter and thus create early infestations on the new crops. Insect control of soil-inhabiting pests and diseases may be achieved by chemical drenches, sprays or granular applications, though some of these adversely affect beneficial insects through fumigant action (or by their systemic activity).

Control of thrips and the tomato leaf-miner (*Liriomyza bryoniae*) can be achieved by spraying a mixture of polybutene plus deltamethrin onto the greenhouse floor. This mixture will kill pests that migrate to the ground to pupate for up to 15 weeks and has the advantage that it leaves the beneficial insects unaffected on the foliage of the plants.

On cucumbers where *P. persimilis* is used to control red spider mite, the tops of the plants often become suddenly and severely infested in mid-summer. During spells of hot weather, the pest migrates upwards but the predator, which is not tolerant of high temperatures, cannot, so an imbalance occurs. It can often be corrected without detriment to the predators by an application of a relatively safe chemical (e.g. tetradifon, dicofol) along the tops of the plants.

Similarly, adult whitefly live mainly on the tops of plants while the parasites and larvae suitable for parasitism are lower down on older leaves. Any sudden excess of adult whiteflies can be safely controlled by spraying the young foliage with a chemical of short persistence (e.g. pyrethrum/resmethrin mixtures).

The chrysanthemum leaf-miner (*Chromatomyia* (=*Phytomyza*) *syngenesiae*) feeds and lays eggs on the upper leaf surfaces, especially at the tops of the plants, while it is parasitized during the larval stages which occur lower down on the older leaves. An excess of adults can be reduced by a mist application of diazinon or pyrazophos over the tops of the plants. Alternatively, a quicker method is to use a thermal fogging machine, from which most of the chemical is deposited on the upper leaf surfaces.

Misting or fogging to control leaf-miner preserves the interaction on the undersides of the leaves between spider mites and predators, and aphids and parasites. However, this selectivity is lost when the plant forms flowering buds as these other pests then migrate to the tops of the plants.

REFERENCES

Hussey, N.W. & Helyer, N.L. (1981) Development of a management programme to ensure whitefly-free early-sown tomatoes on Guernsey. *Rep. Glasshouse Crops Res. Inst.* **1981**: 102–104.

Schulten, G.G.M. & van de Klashorst, G. (1979) Genetics of resistance to parathion and demeton-s-methyl in *P. persimilis* A.H. (Acari: Phytoseiidae). pp. 519–534. In *Proceedings of the 4th International Congress of Acarology, Budapest* 752 pp.

Scopes, N.E.A. & Biggerstaff, S.B. (1977) The use of a temperature integrator to predict the development period of the parasite *Aphidius matricariae. J. Appl. Ecol.* **14**: 799–802.

Voroshilov, N.V. (1979) Heat resistant lines of the mites *P. persimilis* A-H. *Genetika* **15** (1): 70–76.

6. INTEGRATED PROGRAMMES FOR SPECIFIC CROPS

6.1 TOMATOES
J. Woets

6.1a **INTRODUCTION** The possibilities of integrated crop protection in greenhouse crops are largely determined by the availability of biological control methods for the main pests and by the pesticides required to control diseases and minor pests. The changing availability of pesticides is less important than the changing cultural practices such as the decreasing temperature regime, growing in rock-wool and covering the soil with plastic sheets.

Tomato is an important greenhouse crop in Western Europe and has a simple programme of crop protection procedures (Table 6). Sterilization of the soil by steaming or methyl bromide application preceding planting during winter (December–March) is the major control measure as it eliminates soilborne diseases such as tomato mosaic virus (TMV), *Fusarium*, *Verticillium* and soil-borne pests, such as *Meloidogyne*, tomato moth (*Lacanobia oleracea*), tomato leaf-miner (*Liriomyza bryoniae*) and American serpentine leaf-miner (*Liriomyza trifolii*). Another aspect of tomato-growing is the general use in Western Europe of resistant genes against TMV, *Cladosporium* and *Fusarium*. Several varieties also have genes conferring tolerance to *Verticillium* while a few tolerate root-knot nematodes. Only the foliage pests and *Botrytis* require frequent use of direct control measures.

When a tomato variety does not include TMV-resistance, a type of biological control developed by Rast (1975) can be applied. In the seedling stage, tomato plants are inoculated with a mild strain of the TMV virus. This is still practical for a 'beef' tomato, e.g. 'Dombo' and also in several countries, such as Spain and Canada where commercial varieties do not include TMV resistance. By this method, the plants are not completely protected from the pathogenic strain of TMV, but its effect is reduced.

This account is based largely on the situation in the UK and the Netherlands because most biological pest control occurs in these countries (van Lenteren *et al.*, 1980), but aspects in other countries are not neglected.

TABLE 6 List of the common pests and diseases and their recommended control agents in an integrated pest management programme for soil grown tomatoes planted in winter in the UK and the Netherlands.

Pests/Diseases	Recommended agents in UK	Recommended agents in Netherlands
Greenhouse whitefly	*Encarsia formosa* (*Verticillium lecanii*)	*Encarsia formosa*
Greenhouse red spider mite	*Phytoseiulus persimilis*	fenbutatin oxide
Tomato leaf-miner	parasites	parasites, mainly natural
American serpentine leaf-miner	—	parasites, natural control
Peach-potato aphid, Potato aphid	pirimicarb	pirimicarb
Tomato moth, Tomato looper	*Bacillus thuringiensis*	*Bacillus thuringiensis*
Botrytis cinerea	iprodione, vinclozolin, dichlofluanid	iprodione, vinclozolin, dichlofluanid

When tomatoes are grown in the soil, no soil-borne pests can overwinter. The only pests which can be carried over are the greenhouse red spider mite (*Tetranychus urticae* Koch) and – in the Netherlands – the tomato looper (*Chrysodeixis chalcites* Espes) because they overwinter in the structure.

All growers attempt to plant young plants free of pests. This can be achieved either by application of oxamyl granules (UK) or frequent spraying of oxamyl on the seedling benches (Netherlands). To prevent early spider mite infestation, good control during the preceding crop by predators or chemicals is necessary. This is also important because of the reduced effectiveness of the predator during the first months of the year (low air humidity).

6.1b **TOMATO PESTS** These will now be discussed in order of importance:

1. GREENHOUSE WHITEFLY is the key pest in integrated control in tomatoes. Most growers receive a very low infestation on the young plants from the propagation nursery. When the first adult whiteflies have been seen during the cultural practices of planting, tying and watering, the grower requests the dealer in biological control to deliver the first introduction of *Encarsia formosa*. In practice, there are several possibilities facing the grower depending upon how the crop has been infested with whitefly and Figure 32 illustrates how to make rational decisions using a flow chart.

The introduction is repeated at least 2–3 times at 10- or 14-day intervals until black, parasitized scales appear on the crop plants. This common system is called the 'multiple introduction method' or 'dribble method' (Griffin & Savage, 1983) and was presented at the same time by

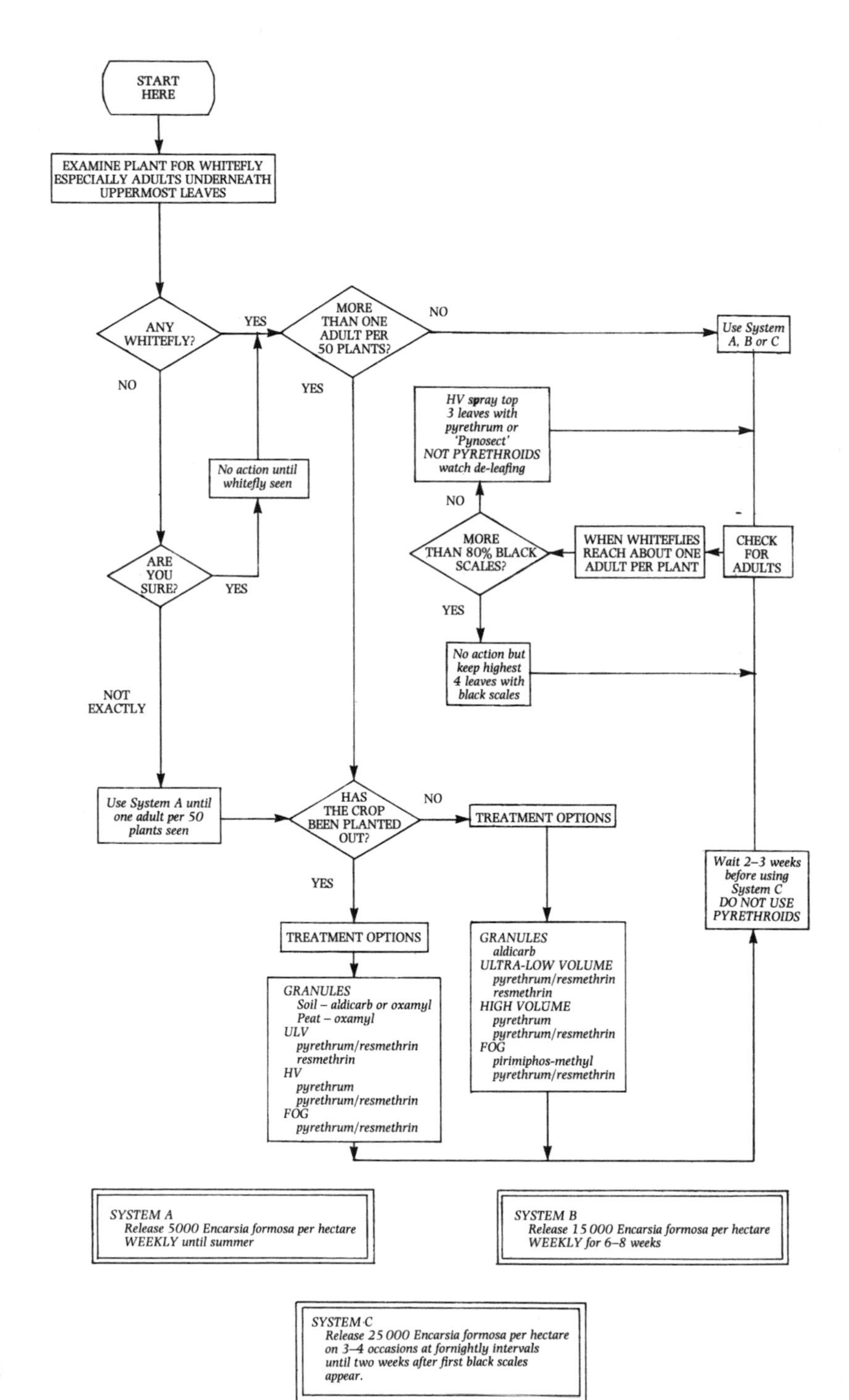
START HERE
EXAMINE PLANT FOR WHITEFLY ESPECIALLY ADULTS UNDERNEATH UPPERMOST LEAVES
ANY WHITEFLY?
YES
NO
MORE THAN ONE ADULT PER 50 PLANTS?
NO
YES
Use System A, B or C
HV spray top 3 leaves with pyrethrum or 'Pynosect' NOT PYRETHROIDS watch de-leafing
No action until whitefly seen
NO
MORE THAN 80% BLACK SCALES?
WHEN WHITEFLIES REACH ABOUT ONE ADULT PER PLANT
CHECK FOR ADULTS
ARE YOU SURE?
YES
YES
No action but keep highest 4 leaves with black scales
NOT EXACTLY
Use System A until one adult per 50 plants seen
HAS THE CROP BEEN PLANTED OUT?
NO
TREATMENT OPTIONS
Wait 2–3 weeks before using System C DO NOT USE PYRETHROIDS
YES
TREATMENT OPTIONS
GRANULES Soil – aldicarb or oxamyl Peat – oxamyl ULV pyrethrum/resmethrin resmethrin HV pyrethrum pyrethrum/resmethrin FOG pyrethrum/resmethrin
GRANULES aldicarb ULTRA-LOW VOLUME pyrethrum/resmethrin resmethrin HIGH VOLUME pyrethrum pyrethrum/resmethrin FOG pirimiphos-methyl pyrethrum/resmethrin
SYSTEM A Release 5000 Encarsia formosa per hectare WEEKLY until summer
SYSTEM B Release 15 000 Encarsia formosa per hectare WEEKLY for 6–8 weeks
SYSTEM C Release 25 000 Encarsia formosa per hectare on 3–4 occasions at fornightly intervals until two weeks after first black scales appear.

Parr (1973), Parr *et al.* (1976) and Woets (1973 & 1978). The initial infestation of whitefly is usually less than 1/100 plants. The *total* number of *Encarsia* introduced as black pupae per m² is 3–4. This technique is common in many countries where biological control is practised – West Germany, Belgium, France, Norway, Sweden and Canada.

Other methods of introduction are the 'pest-in-first' method and the 'banker plant' method (Glasshouse Crops Research Institute, 1976) but they are rarely accepted commercially, mainly for psychological reasons against pest introduction.

Higher multiplication rates of greenhouse whitefly occur on beef tomato varieties than are common on other varieties (Woets *et al.*, 1983b). At least 5 weekly introductions of 1 *Encarsia*/m², which is the usual minimum for whitefly control on cucumbers, is therefore made.

In the Netherlands, two supplementary introductions are necessary shortly after 'interplanting' the new crop in June. The young plants are more nutritious and so stimulate a higher multiplication rate in whitefly.

Some guidance is usually given to the grower during the first and second year. In the Netherlands, guidance is included in the price per m². In the UK, as in most countries, the price is offered per 1000 parasites. In the Netherlands, 3–4 introductions cost 0,20 D.Fl. or £0.05 per m².

2. RED SPIDER MITE is not as important a pest as the whitefly. If allowed to diapause, the red spider mite will remain through the winter as diapause females in crevices of the greenhouse. Good pest control during the autumn, either by predators or chemicals, and a soil application in the seedling stage will protect winter-planted tomatoes from early infestations by diapause spider mite females.

French *et al.* (1976) compared introduction methods. They concluded that a predator release after the first spontaneous occurrence of the pest does not always give satisfactory control, apparently because of the rapid and patchy appearance of the colonies founded by overwintered females. This confirms Dutch experience, where chemical control is generally proposed and involves 1–2 applications of fenbutatin oxide. On the other hand, where the 'pest-in-first' is attempted, satisfactory results follow. One reason for the problems in this crop could be the walking behaviour of the predator which is affected by the exudating hairs on the stems and leaves of tomatoes. Much time is taken for 'cleaning' that is not then available for searching.

Due to the low pest pressure of spider mite on tomatoes, chemical control is usually cheaper than predator release after May. The price for predator application is 20 cents/m² in the Netherlands, which is equivalent to £4.70/1000 predators in the UK.

FIGURE 32 Integrated control of tomato pests – whitefly. (As published in *Grower* by L.R. Wardlow, ADAS, Wye, January 1984.)

3. THE TOMATO LEAF-MINER (*Liriomyza bryoniae* Kalt.) is a limiting pest as long as no specific control method is available that does not harm natural enemies. Generally, the pupae are killed when the grower sterilizes the soil before the winter planting. Young plants must be kept free of miners, mainly by drenching or spraying with oxamyl.

Most growers who start with clean plants in a greenhouse that is free of leaf-miners will experience a spontaneous infestation in June, that will be followed by natural control as long as no insecticides are applied (Woets & van der Linden, 1982 & 1983a). Three parasite species are common: *Dacnusa sibirica* Telenga, *Opius pallipes* Wesmael and *Diglyphus isaea* Walker. The last species appears from June onwards and is able to exterminate leaf-miner populations both by parasitization and host-feeding.

The two mentioned braconid endoparasites, *D. sibirica* and *O. pallipes*, can overwinter in the leaf-miner as pupae in the soil. From observations in 1980 and 1981, we know that overwintering leaf-miner pupae do not always contain parasites. In 1980, both pest and parasites appeared on young tomatoes after overwintering from an autumn crop. They achieved excellent natural control. In 1981, only the pest overwintered and parasites came in from outdoors only during May and June. It proved possible to obtain natural control in this way but was obviously risky. Most biological control of leaf-miners is based on the spontaneous occurrence of both pest and parasites which is checked by a sampling service conducted by the producer of natural enemies (the Netherlands). Most tomatoes are now grown in rock-wool or peat-bags. Tomato leaf-miner pupae remain from the previous autumn and as soon as the greenhouse is heated the miners are stimulated to emerge. When parasites are found in a sample of 50 full-grown leaf-miner maggots, control will be satisfactory. When no parasite larvae are found, an introduction of parasites will be necessary in the generation 5 weeks later. For this purpose *O. pallipes* was produced on a small scale in the Netherlands during 1983 because this species is more reliable for greenhouse use than *D. sibirica* which had been produced for some years. In the UK, *D. isaea* was used by some growers. There is much experience of effective control with *Dacnusa* and *Diglyphus*, both as natural and planned biological control (Wardlow, 1984).

4. THE AMERICAN SERPENTINE LEAF-MINER (*Liriomyza trifolii* Burges) has occurred in greenhouses in Western Europe since the late seventies. It is upsetting the progress made in biological control of the tomato leaf-miner since it occurs in such important vegetable crops as tomato, sweet pepper and aubergine. This species is a native of Florida and, because of excessive insecticide use in celery and chrysanthemum, it is tolerant to many common insecticides. Chemical control is still possible, but only a

few pesticides provide good results with high costs of ingredients and labour.

The parasites of the common tomato leaf-miner have different effects on the exotic species. *Dacnusa* will oviposit in it, though control is poor, while the eggs of *O. pallipes* become encapsulated. *Diglyphus* is doing well but apparently only from June onwards. Its effect is excellent and can lead to complete extermination. So we now need to find out which parasites can be mass-produced most easily and used to control both leaf-miners.

The tomato leaf-miner may occur throughout the whole year outside tomato-houses where soil sterilization is banned. Because of the rapid multiplication rate of American serpentine leaf-miner, and because it is very uncommon on tomato during the winter, it is desirable to identify a natural enemy that will multiply on *L. bryoniae* but which will attack and control the American species during the summer.

Initial experience in North America and Western Europe suggests that there are three promising candidates: the European *Diglyphus isaea* Woelur, the Californian *Chrysocharis parksi* Crawford and the American *Opius dimidiatus* Ashmead.

5. THE PEACH-POTATO APHID (*Myzus persicae*) AND THE POTATO APHID (*Macrosiphum euphorbiae*) are two aphids which occur frequently on tomato. Pirimicarb is a selective aphicide which provides a good control of both without harming parasites such as *Encarsia*. During the summer, spontaneous parasitization of the aphids occurs so that aphicide applications are rarely necessary or even desirable as the elimination of the aphids might exterminate the 'free' parasites.

6. THE TOMATO MOTH (*Lacanobia oleracea*) was originally of no importance, but since peat-bags and rock-wool became popular there has been an increasing carry-over of tomato moth pupae into the next season. Foster (1980) concluded from this survey that the greatest danger is posed by the first generation. However, control is then very easy by spraying with *Bacillus thuringiensis*. However, growers often think the numbers negligible until they observe the second caterpillar generation which can destroy many fruit before control is feasible.

7. THE TOMATO LOOPER (*Chrysodeixis chalcites*) became established in Western Europe during the warm summer of 1976. Several adults were caught in the Netherlands in light traps and it became clear that this invasion initiated a new greenhouse pest. Sweet pepper, aubergine and cucumber are also attacked. *Bacillus thuringiensis* has proved to be an excellent control agent.

8. *Botrytis cinerea* is a very common fungus which is being favoured by changes in the use of energy for climatic control. Fortunately most *Botrytis* fungicides do not harm predators and parasites.

6.1c **GENERAL COMMENTS** Growers do not like to spray and so there is a general custom to dust, smoke or drench as much as possible. Since negative experience with the use of *Encarsia* in 1983, it became clear through an intensive survey of the beneficial producers that frequent dusting of chemicals decreases the efficiency of both *Encarsia formosa* and leaf-miner parasites. It is, therefore, now advised to apply dust only when there is a strong balance between pest and parasites.

The change to alternative substrates includes several aspects that interfere with pest control. Pupae of both tomato moth and leaf-miner may overwinter. But there are also some advantages in that parasites may also overwinter. The new watering techniques (dripping) provide new application possibilities and these are now often used for the application of chemicals such as oxamyl.

Also, the plastic cover over the soil and the plastic screens over the crop influence plant protection. In the past, dusting of malathion was used against the tomato leaf-miner on young plants. After some weeks, the plants and heating pipes were cleaned by spraying water. Then *Encarsia* could be released and worked effectively. Since 1983, it is clear that this is no longer possible under the new conditions. The malathion maintains its vapour effect much longer on the dry plastic covers. An extra effect is the reduced ventilation in modern houses due to thermal screens and insulated walls. It is possible that the new synthetic materials have negative vapour effects on parasites as a parallel to the 'softeners' of water hoses and the hoses connecting heating pipes, which can have striking phytotoxic effects on plants.

Another important use of these polythene sheets is for the application of Thripstick® (a commercial product based on polybutenes and deltamethrin) which has been shown experimentally to provide very effective control of *Liriomyza* spp. as the larvae fall to the ground to pupate.

When the balance between pest and beneficial is incorrect, it may be worthwhile to restore the balance. For that purpose, control agents which do not harm beneficials are needed. In this way *Verticillium lecanii* as Mycotal® (produced by Tate & Lyle, GB), bioresmethrin and Safer's Insecticidal Soap® from Canada are important. When such chemicals are available, it is easier to stimulate growers to change to biological control, because there is something in reserve for emergency in cases of poor control. A flow chart rationalizing some of these decisions (Figure 33)

FIGURE 33 Integrated control of tomato pests other than whitefly. (As published in *Grower* by L.R. Wardlow, ADAS, Wye, January 1984.)

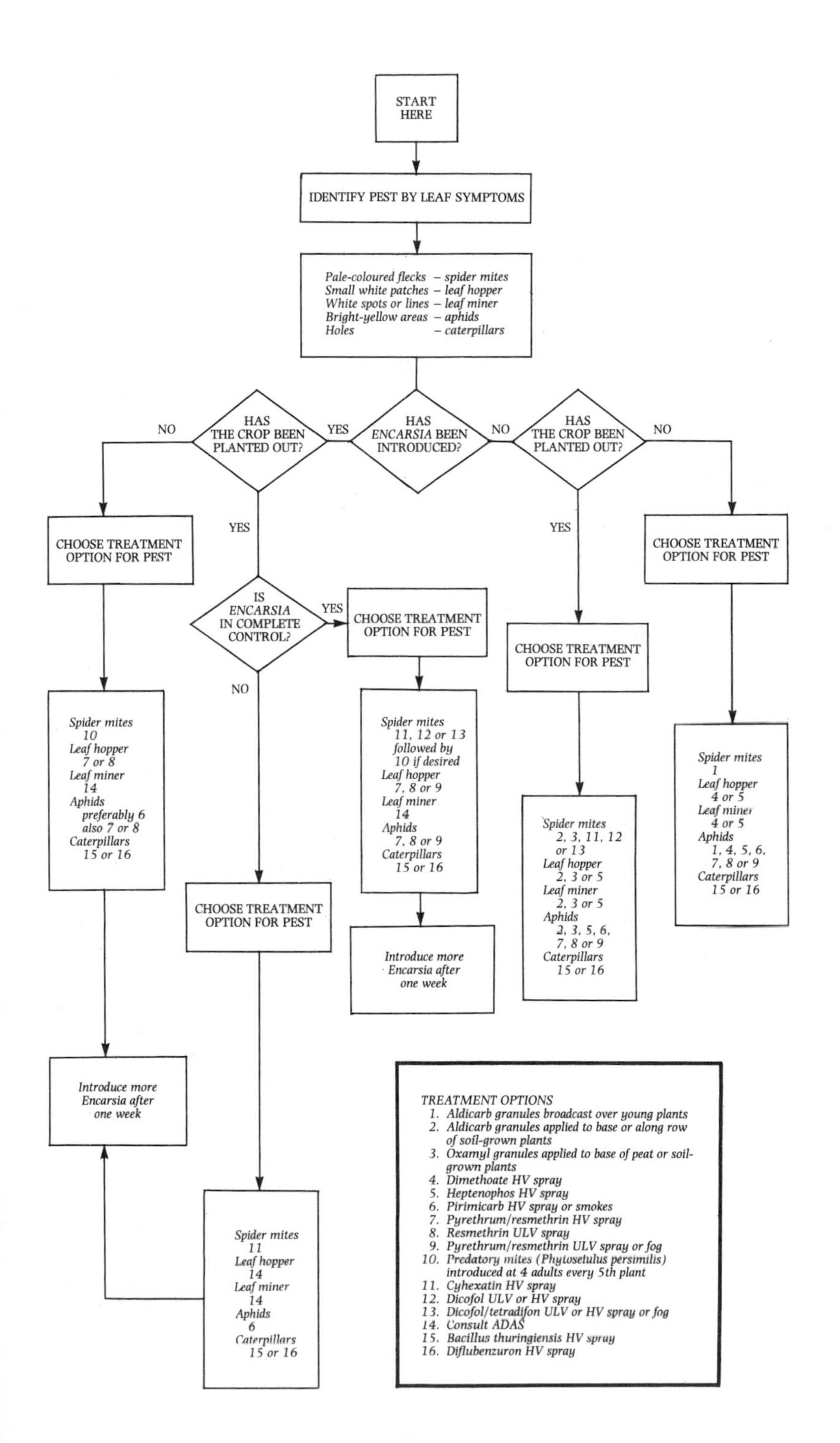
START HERE
IDENTIFY PEST BY LEAF SYMPTOMS
Pale-coloured flecks – spider mites
Small white patches – leaf hopper
White spots or lines – leaf miner
Bright-yellow areas – aphids
Holes – caterpillars
HAS THE CROP BEEN PLANTED OUT?
NO
YES
HAS ENCARSIA BEEN INTRODUCED?
NO
HAS THE CROP BEEN PLANTED OUT?
NO
YES
CHOOSE TREATMENT OPTION FOR PEST
YES
IS ENCARSIA IN COMPLETE CONTROL?
YES
CHOOSE TREATMENT OPTION FOR PEST
NO
CHOOSE TREATMENT OPTION FOR PEST
CHOOSE TREATMENT OPTION FOR PEST
Spider mites 10
Leaf hopper 7 or 8
Leaf miner 14
Aphids preferably 6 also 7 or 8
Caterpillars 15 or 16
Spider mites 11, 12 or 13 followed by 10 if desired
Leaf hopper 7, 8 or 9
Leaf miner 14
Aphids 7, 8 or 9
Caterpillars 15 or 16
Spider mites 2, 3, 11, 12 or 13
Leaf hopper 2, 3 or 5
Leaf miner 2, 3 or 5
Aphids 2, 3, 5, 6, 7, 8 or 9
Caterpillars 15 or 16
Spider mites 1
Leaf hopper 4 or 5
Leaf miner 4 or 5
Aphids 1, 4, 5, 6, 7, 8 or 9
Caterpillars 15 or 16
CHOOSE TREATMENT OPTION FOR PEST
Introduce more Encarsia after one week
Introduce more Encarsia after one week
Spider mites 11
Leaf hopper 14
Leaf miner 14
Aphids 6
Caterpillars 15 or 16
TREATMENT OPTIONS
1. Aldicarb granules broadcast over young plants
2. Aldicarb granules applied to base or along row of soil-grown plants
3. Oxamyl granules applied to base of peat or soil-grown plants
4. Dimethoate HV spray
5. Heptenophos HV spray
6. Pirimicarb HV spray or smokes
7. Pyrethrum/resmethrin HV spray
8. Resmethrin ULV spray
9. Pyrethrum/resmethrin ULV spray or fog
10. Predatory mites (Phytoseiulus persimilis) introduced at 4 adults every 5th plant
11. Cyhexatin HV spray
12. Dicofol ULV or HV spray
13. Dicofol/tetradifon ULV or HV spray or fog
14. Consult ADAS
15. Bacillus thuringiensis HV spray
16. Diflubenzuron HV spray

will be found helpful and can, of course, be modified in the light of local experience.

REFERENCES

Foster, G.N. (1980) Possibilities for the control of tomato moth (*Lacanobia oleracea*). *Bull. IOBC/WPRS Working Group on Integrated Control in Glasshouses, Vantaa* **3** (3): 45–52.

French, N., Parr, W.J., Gould, H.J., Williams, J.J. & Simmonds, S.P. (1976) Development of biological methods for the control of *Tetranychus urticae* on tomatoes using *Phytoseiulus persimilis*. *Ann. Appl. Biol.* **83**: 177–189.

Glasshouse Crops Research Institute (1976) The biological control of tomato pests. *Glasshouse Crops Res. Inst. Growers' Bull.* No. 3: 23 pp.

Griffin, M.J. & Savage, M.J. (1983) Control of pests and diseases of protected crops 1983. Tomatoes. MAFF, ADAS Booklet No. 2243: 110 pp.

Parr, W.J. (1973) The use of integrated control in glasshouses in Great Britain. *Bull. IOBC/WPRS Working Group on Integrated Control in Glasshouses, Littlehampton* **1973** (4): 32–35.

Parr, W.J., Gould, H.J., Jessop, N.H. & Ludlam, F.A.B. (1976) Progress towards a biological control programme for glasshouse whitefly (*Trialeurodes vaporariorum*) on tomatoes. *Ann. Appl. Biol.* **83**: 349–363.

Rast, A.Th.B. (1975) *Variability of Tobacco Mosaic Virus in Relation to Control of Tomato Mosaic Virus in Glasshouse Tomato Crops by Resistance Breeding and Cross Protection* Pudoc, Wageningen. 76 pp.

van Lenteren, J.C., Ramakers, P.M.J. & Woets, J. (1980) World situation of biological control in greenhouses with special attention to factors limiting application. *Meden. Rijksfac. Landb. Gent* **45** (3): 537–544.

Wardlow, L.R. (1984) Monitoring the activity of tomato leaf-miner (*Liriomyza bryonia*) and its parasites in commercial glasshouses in Southern England. *Meded. Rijksfac. Landb. Gent* **49** (In press).

Woets, J. (1973) Integrated control in vegetables under glass in the Netherlands. *Bull. IOBC/WPRS Working Group on Integrated Control in Glasshouses, Littlehampton* **1973** (4): 32–35.

Woets, J. (1978) Development of an introduction scheme for *Encarsia formosa* Gahan (Hymenoptera: Aphelinidae) in greenhouse tomatoes to control the greenhouse whitefly, *Trialeurodes vaporariorum* (Westwood) (Homoptera: Aleyrodidae). *Meded. Rijksfac. Landb. Gent* **43** (2): 379–385.

Woets, J. & van der Linden, A. (1982) On the occurrence of *Opius pallipes* Wesmael and *Dacnusa sibirica* Telenga (Braconidae) in cases of natural control of the tomato leaf-miner *Liriomyza bryoniae* Kalt. (Agromyzidae) in some large greenhouses in the Netherlands. *Meded. Rijksfac. Landb. Gent* **47** (2): 533–540.

Woets, J. & van der Linden, A. (1983a) Observations on *Opius pallipes* Wesmael (Hym., Braconidae) as a potential candidate for biological control of the tomato leaf-miner *Liriomyza bryoniae* Kalt. (Dipt., Agromyzidae) in Dutch greenhouse tomatoes. *Bull. IOBC/WPRS Working Group on Integrated Control in Glasshouses, Darmstadt* **6** (3): 134–141.

Woets, J. & van der Linden, A. (1983b) Biological control of the greenhouse whitefly. *A. Rep. Glasshouse Crops Res. Exp. Stn, Naaldwijk* **1982**: p. 103.

6.2 CUCUMBERS

N.W. Hussey

Cucumbers were the first crop on which an integrated programme was successfully developed in order to accommodate biological control of spider mites. This pest had developed so many resistant strains that chemical control had almost completely failed.

It was immediately obvious that whiteflies occupied the same niche on the plant as mites and their predators, while tests revealed that the pesticides normally used for whitefly control were lethal to *Phytoseiulus*. It was, therefore, necessary to utilize biological control for whiteflies as well. Since *Encarsia* had been produced and used in the UK since 1929, most of the initial research took place there. Readers will recall that these pioneering studies were based on the 'pest-in-first' concept so that the levels of control achieved were commercially satisfactory from the beginning. From those nurseries where large-scale trials were conducted there was an enthusiastic demand for the techniques to be adopted as a routine. Parasites and predators were therefore provided by small rearing-units supplemented by the GCRI at Littlehampton. This preliminary commercial experience rapidly showed that certain other pests then assumed major status since their importance had hitherto gone unrecognized because they had formerly been readily controlled by the routine sprays used for spider mite and whitefly.

The most serious of these 'minor' pests proved to be *Thrips tabaci*. Though these insects were readily killed by pesticides, it was not possible to use a sufficiently selective material to harmonize with natural enemies which were active on the leaves on which thrips were breeding. However, as pupation takes place on the ground, the strategy of applying a persistent pesticide to the soil surface was immediately effective. In the 1960s, cucumbers were grown on manure beds and the plants trained along wires below the low roofs. We now know that these conditions, which led to wet, or at least very damp, floors, encouraged epizootics of the fungal parasites, *Entomophthora thripidum* and *E. parvispora*. The rates of population increase were, therefore, depressed although this was not appreciated at the time. The use of γ-HCH provided effective control if applied at the first sign of damage. This technique operated successfully until the mid-1970s when cucumbers began to be planted earlier in the year. Where traditional planting was employed in March, outside air temperatures were rising when thrips required treatment so that γ-HCH vapour concentrations were diluted by the daily ventilation of the houses. On the other hand, planting in mid-winter involved long periods without ventilation so that γ-HCH vapour accumulated and predators were rapidly killed. A new approach was therefore required which did not have associated vapour problems. The development of Thripstick® (equal parts of polybutene 5 and water plus 4.2 cm^3 deltamethrin/l) provided a novel and effective answer, though registration and marketing problems have delayed its implementation in many countries. This material is particularly appropriate where hydroponic or peat-bolster techniques are used for growing cucumbers as they are placed on plastic sheeting to which Thripstick® can be conveniently applied.

Another approach to thrips control is biological and the Dutch have successfully developed the predatory mites, *Amblyseius cucumeris* and *A. mackenziei*, though it seems that these predators are able to deal efficiently with the pest only on manure beds where thrip population increase is slow because, in drier conditions, they cannot respond sufficiently rapidly to prevent severe damage.

Strains of *Verticillium lecanii* could also provide control if the market were sufficiently large to justify commercial development.

There are evidently a number of selective approaches to this problem and it is clear that much of the reluctance of Dutch growers to adopt an integrated approach is due to the widespread use of tetrachlorvinphos there. This pesticide has deleterious effects on natural enemies, even at low dosages. The use of cyanide for whitefly control in the Netherlands, which is not permitted in other countries, further simplifies their pest problem.

Another minor pest which has almost disappeared in the face of artificial cultural substrates is the mite, *Tyrophagus longior* ('French fly'), a glistening white, globular mite with long bristles which caused considerable problems where young cucumbers were planted into manure or straw bales (Plate 27). Both substrates contain these mites which feed on moulds in the decaying organic matter. Although not phytophagous, they often swarm up onto the young plants, eating minute holes in the youngest leaves which then become quite large as the leaves expand. More seriously they may cause a proportion of the infested plants to go 'blind'. Traditionally, they were controlled by drenches and sprays of parathion. This material was toxic to predators and, being partially systemic, prevented use of the predator for several weeks. Later, sprays of pirimiphos-methyl (25 g/100 l) provided an effective and less persistent alternative.

Another interesting and unexpected pest was the mite, *Bryobia praetiosa*. This species often invades cucumbers where the houses are surrounded by rank vegetation in which these clover mites abound. Unexpectedly, it was found that the fungicide, Elvaron® (dichlofluanid), provided effective control without harm to natural enemies.

Hence, in mounting an integrated programme in the UK, all the essential selective elements were available and the recommended programme takes the following form:

1. *Tyrophagus* MITES If the crop is to be raised by traditional culture in manure or on straw bales, the substrate should be treated with pirimiphos-methyl (25 g/100 l) before planting. Alternatively, a mushroom compost additive, such as Sporavite®, which contains sources of both carbohydrate and nitrogen should be used to alter the microbial interactions involved in the degradation of organic matter, so that bacterial populations dominate those of the fungi. This technique has been shown to drastically reduce the incidence of 'French fly' which is normally expected under these conditions. Where damage by this pest does occur, pirimiphos-methyl must be applied to the crop and the introduction of natural enemies of other pests deferred for at least 3 weeks.

2. RED SPIDER MITE is usually found to be hibernating in vast numbers in the house structure when the programme is first attempted. This results in a severe infestation of diapausing females within a few days of planting. Predators feed reluctantly on these 'red' mites, reproducing only slowly. The crop must, therefore, be artificially infested by placing 10 green, actively feeding mites on every plant and allowing them to reproduce until the mean leaf damage index reaches 0.4, when *Phytoseiulus* is introduced. The Dutch supplier, Koppert BV, supplies the predator together with spider mites in a 'carrier' which can be conveniently shaken over the crop, so providing food to ensure a rapid preliminary build-up of *Phytoseiulus* to deal with mites emerging from their hibernation site. Control should be maintained for several months but in Northern Europe particular care should be taken to ensure that no mites remain by August. Any mites remaining active at this time will enter diapause in response to decreasing day length and so perpetuate the mite problem the following year. Where this has been avoided, the introduction of 8 predators/m² should suffice to prevent problems. It is important to recognize that the cyanide commonly used in the Netherlands to control whiteflies is partially toxic to active stages of the predator and so will have deleterious effects on spider mite control if the population of *Phytoseiulus* is low at the time of treatment. Further, in hot summer weather, the predator may desert the young apical cucumber growth which must be protected by sprays of cyhexatin or petroleum oils.

3. WHITEFLY control by *Encarsia* is recognized as being less effective on cucumbers than on tomatoes despite the higher growing temperatures. This is largely caused by: (a) glandular hairs on the leaf, which reduce mobility of the parasite, and (b) the high reproductive rate achieved by whitefly on this host. Since the key to successful biological control is the prevention of dense populations of whitefly on small groups of plants, the

planned introduction of the pest, at 10 whiteflies/5 plants within 2 weeks of planting, is strongly recommended, followed by 3 introductions of *Encarsia* at the rate of 10/plant to the sites of pest introduction at intervals of 14 days. Especial care must be taken to avoid infestation during propagation, either by the use of oxamyl or by the introduction of *Encarsia* by the dribble method. Five introductions of 2 parasites/m² should be made at weekly intervals commencing as soon as the first 3rd-instar host scales appear on the crop.

Occasionally some imbalance may occur on certain plants and it is important to prevent excessive honeydew deposit, which will occur when more than 30 adults/leaf occur on the upper leaves. These should be contained by applying quinomethionate, natural pyrethrum or Pynosect® (a mixture of bioresmethrin and resmethrin) to the affected plants. The recent development of *Verticillium* as the commercial product, Mycotal®, provides an excellent selective technique as it sets up an epizootic among the whiteflies if it is applied when the pest occurs on 10% of the plants. The fungus is totally compatible with *Encarsia* as it does not affect the adults though it may kill some parasitized scales.

It is important to inspect all the foliage being 'trimmed-off' the plants and if any leaves carry large numbers of unemerged parasites they should be kept in the glasshouse for a further 10–12 days.

Research in Belgium and the Mediterranean, especially Sicily, has confirmed the value of yellow traps placed at 1/5 m² (3 m intervals in alternate rows). The traps are plastic plates about 20 cm in diameter and are coated with a glue, Temocid®, which, in hot Mediterranean conditions, must be reapplied every 15–20 days. These traps must be introduced into the crop at planting. If necessary, they may be supplemented by quinomethionate sprays (10 g/100 l), which are sometimes needed before *Encarsia* is released if whitefly populations are high.

4. THRIPS In traditional culture, it is only necessary to keep the earth floor of the glasshouse permanently damp so as to encourage *Entomophthora* sp. to attack the pre-pupae. About 10 *Amblyseius mckenziei* should be introduced to each plant when the characteristic feeding marks of thrips first appear.

Where cucumbers are grown in peat-bolsters, rock-wool or nutrient-film systems, Thripstick® should be applied at 300 l/ha to the plastic sheets and bag surfaces directly damage is detected. This treatment may have to be repeated after 10 weeks.

5. APHIDS When experiments on integrated control first began, much difficulty was experienced with the cotton aphid (*Aphis gossypii*), which multiplies at a rate of more than 10 × per week. Attempts to use exotic

parasites, coccinellids and chrysopids all failed, though the midge, *Aphidoletes aphidimyza*, showed some promise. However, the advent of pirimicarb (25 g/100 l) provided an effective and selective control which must be applied at the first sign of damage. *A. gossypii* is a small globular aphid with very short cornicles. When not crowded, it is black or bottle green but, as population density rises, it may become pale yellow. Large numbers cause the collapse of older leaves and severe distortion of younger foliage. The maximum population which can be tolerated without loss of yield is 7 aphids/cm^2 of leaf area.

A large green aphid, *Macrosiphum euphorbiae*, may occasionally develop on isolated groups of plants where light green patches develop on the leaves, with severe puckering of the leaf tissue, even when there are less than 10 aphids/leaf. Spot treatment with nicotine (112 g/100 l) is the recommended control.

6. LEAF-MINERS Occasionally *Liriomyza trifolii* and/or *L. bryoniae* cause problems on cucumbers but experience in the South of Europe suggests that, if yellow traps (see *Whitefly* above) are suspended just above the apical foliage at the first sign of attack, then invasion of the crop by *Diglyphus iseae* from outside will contain the problem without further treatment.

7. FUNGI The principal foliar disease is cucumber mildew (*Sphaerotheca fuligineae*) for which the fungicide, benomyl, should *not* be used, as it has been demonstrated that this fungicide induces sterility in the predatory mite, *Phytoseiulus*, though there is no obvious direct toxicity. Mildew is discouraged by hot, damp growing conditions but under modern cultural methods the dry air demands routine use of fungicides. The following are the most selective for widespread use: bupirimate (50 g/100 l), imazalil (10 g/100 l), carbendazim (25 g/100 l) or cupric ammonium carbonate (21 g/100 l). In damp conditions, *Didymella bryoniae* may cause lesions on fruit and leaves but can be controlled by chlorothanolil (110 g/100 l) or iprodione (50 g/100 l).

6.3 CHRYSANTHEMUMS
L.R. Wardlow

6.3a **EARLY EXPERIMENTAL WORK ON INTEGRATED CONTROL** One way of overcoming the resistance to acaricides developed by two-spotted or red spider mite is to use the predatory mite, *Phytoseiulus persimilis*. Scopes & Biggerstaff (1973) tried one method of inoculating year-round chrysanthemum cuttings with both pest and predator prior to planting, this worked well but has not been adopted commercially for two main reasons: (a) nurserymen are reluctant to infest their crops with pests, and (b) cuttings may be contaminated with pesticide residues that kill predators.

Another approach to integrated control was described by Crosse (1980) in which predators were introduced onto plants already treated with aldicarb. This was successful when predators were introduced 3–4 weeks after the aldicarb treatment, but the rate of introduction (10 adult *P. persimilis*/10 plants) is too expensive for commercial use at current costs. Lower rates of predator introduction are unlikely to work consistently.

In 1978, an investigation began in Kent, UK, into a post-planting system in which *Phytoseiulus persimilis*, leaf-miner parasites, *Verticillium lecanii* for aphids and *Bacillus thuringiensis* for caterpillars were integrated with short-persistence pesticides for the control of other pests. Since that time ADAS entomologists in several regions of England have co-operated with colleagues at the GCRI and commercial breeders of natural enemies to find a workable commercial programme. The development of such a programme relied on the commercial availability of both leaf-miner parasites and the aphid fungal disease, *Verticillium lecanii* (Hall, 1980).

6.3b **SOME PRINCIPLES AND COMPONENTS OF THE PROGRAMME** MITES Before 1980, predators of two-spotted spider mite were more than twice as expensive as they are today, so a lower introduction rate was required. A minimum number of introduction sites was also necessary to reduce the cost of distributing predators within the nursery. In the ADAS trials, it was found that 10 adult predators/100 plants resulted in good control, particularly where spider mites were relatively numerous. Problems seemed most likely to arise when predators died quickly because the plants were not sufficiently infested with mites. In spite of this, higher rates of introduction, say 30 adult predators/100 plants, are recommended by one commercial supplier in order to reduce the risks of failure. Most nurserymen use higher rates on cultivars such as 'Pink Gin', 'Romark' and 'Yellow Snowdon' that are especially prone to damage by mites. I think it important not to 'dilute' the introduction of predators below 1 site/100 plants as they will not spread sufficiently through the dense mass of chrysanthemum foliage.

One supplier has developed a system of incorporating a known number of predators and spider mites for food with bran flakes in a plastic 'shaker' bottle (Figure 39). The grower simply broadcasts the flakes and predators over the crop, ensuring a wide distribution.

Predators may be introduced 3 or 4 weeks after planting; growers often choose the earlier timing for the rapidly maturing crops grown in the height of summer or for crops grown in blocks or pots. Predators need at least 5 weeks to achieve control of spider mites and even 7 weeks in cool dull conditions. It is important that spider mites are eliminated before the flowers open fully since predators seem not to follow mites readily into the bloom. Most nurserymen have been consistently satisfied with the results of using predators against spider mites on chrysanthemums.

APHIDS The fungal disease of aphids, *Verticillium lecanii*, requires high relative humidities for effective results (Hall, 1980). It is, therefore, particularly suited to the conditions in year-round chrysanthemum greenhouses where blackouts are pulled over the crop each night. *Verticillium lecanii* is particularly effective against peach-potato aphid (*Myzus persicae*) but is sometimes less effective against melon-cotton aphid (*Aphis gossypii*), which often escapes contact with the spores as it moves only limited distances. Recent experience has shown that repeated applications of 'split doses' (one tenth normal) greatly improve control of this species. The shiny brown chrysanthemum aphid (*Macrosiphoniella sanbornii*) is not killed by the fungus as the humidity of the stems on which it feeds is too low and it must be controlled by an HV spray of pirimicarb as soon as it is seen. This pesticide is usually fairly safe to parasites and predators. If too many aphids are present at the time of treatment with *V. lecanii*, the large number of white corpses may subsequently spoil the appearance of the crop. To counter this problem, the fungus is sprayed over the crop 2 weeks after planting when aphid numbers are usually low. The disease takes 2–3 weeks to develop on the aphid but, although aphids continue to breed, their progeny usually also contract the disease. A single treatment is adequate for the control of aphids introduced on cuttings or for low infestations in the first 2 weeks after planting. However, immigrations of aphids from outdoors later in the life of the crop must be treated with a short persistence or 'safe' insecticide.

LEAF-MINERS The chrysanthemum leaf-miner (*Chromatomyia syngenesiae*) invariably occurs on crops where pesticide treatment has been relaxed. Incidence on cuttings has diminished since the fear of *Liriomyza trifolii* becoming established on stock plants has persuaded propagators to improve their pesticide programmes. However, *C. syngenesiae* may enter greenhouses during the summer from outdoors and may attack crops at any stage of growth. An introduction of the endoparasite, *Dacnusa sibirica*, at the rate of 3 adults/1000 plants early in the life of each crop has generally worked well against first generation leaf-miners, but a second introduction of parasites is needed to control the progeny of any survivors. Where possible, the ectoparasite, *Diglyphus isaea*, is preferred for the second introduction but commercial supplies are not consistently available. Immigrations of *D. isaea* from outdoor weeds cannot always be depended upon but they are a welcome bonus when they occur. In the UK, nurserymen use ADAS laboratory facilities for assessment of larval parasitism to determine whether additional releases of up to 1000 parasites/ha are required. One commercial breeder advocates the release of parasites only when the first feeding and oviposition punctures made by adult leaf-miners are seen; 3 introductions of 1250 *Dacnusa sibirica*/ha are then made every 2 weeks.

CATERPILLARS The bacterial pesticide, *Bacillus thuringiensis*, or the chitin-inhibitor, diflubenzuron, kill caterpillars without affecting natural enemies. Most nurserymen wait until the first signs of damage are seen before including treatment in the programme. To obtain satisfactory results, spraying must be thorough, covering as much of the undersides of leaves as possible.

OTHER PESTS Onion thrips (*Thrips tabaci*), common green capsid (*Lygocoris pabulinus*), tarnished plant bug (*Lygus rugulipennis*), greenhouse leaf-hopper (*Hauptidia maroccana*) and the common earwig (*Forficula auricularia*) are at present difficult to control by biological means. Steam sterilization of soil helps to reduce the incidence of these pests but otherwise pesticides such as nicotine, diazinon or carbaryl must be used. Treatments must be timed carefully to avoid serious harm to natural enemies.

6.3b **THE INTEGRATED PROGRAMME** Experimental work has defined a broad-based programme in which pesticides are used only when necessary. This approach should delay or avoid pesticide resistance problems in the future. Since investigational work began, a stock of *Phytoseiulus persimilis* maintained at the ADAS Centre, Reading, UK, has been found to survive treatment by standard doses of diazinon and up to 10 × the normal dose of carbaryl. This prompted Crosse (1980) to produce a simplified programme using these resistant predators for the control of thrips, leaf-miner, capsids and leaf-hoppers; with careful integration of diazinon. This has since been tested successfully in the field. Both the broad-based and simplified programmes are given below (Tables 7 & 8) but it must be emphasized that they are designed only for a typical year-round crop (planted out after 2 weeks propagation on a rooting bench). The programmes are fairly flexible and may be adapted both for timing and treatment to suit either the short-term pot crop or crops planted in peat blocks after 3 weeks on the rooting bench.

6.3c **FURTHER PRACTICAL CONSIDERATIONS WITH INTEGRATED PEST MANAGEMENT** It is important that both pests and natural enemies are monitored weekly on every crop; examination of 20 plants in every 1000 gives a practical estimate of the situation, especially if plants at the ends of beds or others near doorways are also examined. Experience in commercial greenhouses has provided a number of thresholds (listed below) at which prompt action must be taken to prevent the pest/enemy relationship getting out of balance.

Thresholds For each pest, the density at which appropriate remedial action becomes necessary is defined as follows:

APHIDS If aphids are more numerous than 1–2/plant a few days prior to treatment with *Verticillium lecanii*, the crop must immediately be sprayed with pirimicarb or nicotine. The fungal disease can then be applied as usual.

Unless more than 90% of aphids are affected by *Verticillium lecanii* 3 weeks after treatment, an aphicide must be applied immediately. Check

TABLE 7 Broad-based integrated pest control programme for year round chrysanthemums

Weeks after planting	Action	Details
1	Introduce *Dacnusa* spp. parasites against leaf-miner	3 adults/1000 plants
2	(i) *Early in the week* Spray carbaryl, diazinon or nicotine against thrips	Rates as recommended on product labels
	(ii) *Late in the week* Spray *Verticillium lecanii* against aphids	2.5 g/l water (500–1000 l water/ha)
4	Introduce *Phytoseiulus persimilis* against two-spotted spider mite	10 adults/100 plants
4–6	Spray *Bacillus thuringiensis* against caterpillars	1 kg/1000 l water/ha
6	Introduce leaf-miner parasites (preferably *Diglyphus isaea*)	3 adults/1000 plants

TABLE 8 Simplified integrated pest control programme using diazinon-resistant *Phytoseiulus persimilis*

Days after planting	Action	Details
10	Spray diazinon against thrips and leaf-miners	Rate as recommended on product label
14	Spray *Verticillium lecanii* against aphids*	2.5 g/l water (500–1000 l water/ha)
20	Spray diazinon against thrips, leaf-miners etc.	Rate as recommended on product label
28	Introduce diazinon-resistant *Phytoseiulus persimilis* against two-spotted spider mite	10 adults/100 plants
40 – harvest	(i) Spray diazinon against thrips or leaf-miners or nicotine against aphids as necessary.	Rates as recommended on product labels
	(ii) Spray *Bacillus thuringiensis* against caterpillars at first signs of damage	1 kg/1000 l water/ha

*Where glasshouse temperatures or humidities are suitable for using *V. lecanii*, or where nurseries suffer attacks from species of aphids which are not so easily controlled by the fungal disease, regular pirimicarb sprays should be included in the programme. If aphids are resistant to pirimicarb, nicotine sprays should be applied.

that the humidity is sufficiently high in hot dry weather and, if not, consider supplementing with overhead misting.

TWO-SPOTTED SPIDER MITE At the first signs of bud colour, wherever there are more than 5 mites/leaf, adult *Phytoseiulus persimilis* should also be obviously present. Where this is not the case, cyhexatin spray must be applied immediately.

LEAF-MINER Parasitism of larvae should be checked regularly in a laboratory. When less than 90% are parasitized, several weekly introductions of up to 1000 parasites/ha will be required. If this is thought uneconomic, then pyrazophos sprays should be substituted.

CATERPILLARS Where any caterpillars survive treatment, nurserymen should try to improve their spraying technique and apply repeat sprays of *Bacillus thuringiensis* or diflubenzuron. Frequently, caterpillars are difficult to find and the success of control must be measured by changes in the amount of leaf damage.

THRIPS AND CAPSIDS At the first signs of petal damage, apply weekly sprays of diazinon to later crops at the first opening bud stage onwards.

REFERENCES

Crosse, J.V. (1980) The effect of aldicarb on the control of red spider mite, *Tetranychus urticae* (Koch), by the predator *Phytoseiulus persimilis* Athias-Henriot on year-round chrysanthemums. *Pl. Pathol.* **29**: 184–190.

Hall, R.A. (1980) Control of aphids by the fungus, *Verticillium lecanii*: Effect of spore concentration. *Entomologia Exp. Appl.* **27**: 1–5.

Scopes, N.E.A. & Biggerstaff, S.M. (1973) Progress towards integrated pest control on year-round chrysanthemums. pp. 227–234. In *Proceedings of the 7th British Insecticide and Fungicide Conference*, 1089 pp.

Stone, L.E.W., Lane, A., Hammon, R.P., Gould, H.J. & Potter, R.F. (1977) Aldicarb treatment of year-round chrysanthemums for control of red spider mite, *Tetranychus urticae* (Koch), in 1976. *Pl. Pathol.* **26**: 109–111.

6.4 FOLIAGE PLANTS
N.E.A. Scopes

The foliage plant industry provides a wide range of plants for decoration in the home, offices and public buildings (Plate 28). Plants are sold for visual effect and should be totally free of pests. Economics demand rapid production techniques, resulting in soft plants which tend to be susceptible to both pest and pesticide damage. Development of integrated pest management programmes has been initiated slowly, but interest is being shown in *Encarsia formosa* in Florida while an alternative to *Phytoseiulus persimilis, P. macropolis* – a naturally-occurring species – has been evaluated to control *Tetranychus urticae.*

Biological agents are now being used on a limited scale on foliage plants in interior landscapes by contract maintenance companies. Contract companies provide routine maintenance of plants in offices,

FIGURE 34 Treating tall amenity plants with pesticides.

halls and public buildings, where plants are often grown in unsuitable conditions of light and humidity (Figure 34). Pests seem to thrive on ailing plants and chemical remedies are hampered by the fact that only 'safe' materials can be used in public buildings. Any chemicals used should be free of nasty odours, while the problems of spraying in carpeted and glass-windowed halls and offices can be readily appreciated.

It has, until recently, been normal practice to remove and replace smaller infested plants but escalating costs are forcing contracting companies to find ways of prolonging plant life (some have even resorted to silk trees).

Trials of biological control in the USA and UK have given good results and one company in the UK successfully operates a complete integrated pest management programme.

The programme calls for great care in selecting plants for purchase. Any not totally pest-free are rejected. This is of paramount importance with larger plants (more than 2–3 m tall) which are expensive to replace. As with most contractors, plants are bought in direct from nurseries in Europe and, on receipt, are held for acclimatization, allowing further checks for any signs of pests or diseases. A spray with petroleum oil is often applied both to improve leaf quality and as a precaution against pests. Routine introductions of *P. persimilis* are made every 2–3 weeks in the holding house. (Frequently depends on plant-purchasing schedules.)

Once in the interior landscape, plants are inspected regularly, introductions of *P. persimilis* are made every 8 weeks and, if mealybugs (*Planococcus citri*) are found, white oil sprays are immediately applied, followed by introductions of a selection of parasites (*Leptomastix dactylopii*, *Anagyrus pseudococci*, *Leptomastidea abnormis* and *Pauridea peregrina*) which are currently being tested for efficacy at a number of sites. Releases of adult *Cryptolaemus montrouzieri* have been made but total success has not often been achieved.

Scale insects have received little attention in the UK, but suitable parasites are being sought. The use of Thripstick® (deltamethrin mixed into low-viscosity polybutenes) painted onto tree stems has given encouraging results by killing and preventing the spread of both young scales and mealybugs.

Verticillium lecanii has not been used in interior landscapes owing to the prevailing low humidities but it has great potential in production nurseries where high humidities are needed. It may also find use under temporary plastic tents mounted in position for 2 days or so to encourage spore germination.

Encarsia formosa has occasionally been used to control whiteflies (*Trialeurodes vaporariorum*), though this insect has seldom been a problem.

Vine weevils are an occasional problem, especially in soil beds, and trials with entomophilic fungi and nematodes have shown some promise.

Chemical sprays are applied only as necessary, often as spot treatments with a small hand-sprayer. The selection of available materials is very limited, as they must be safe, odour-free, of short persistence and without residual hazards (Section 5.1).

Similar methods are being tested in tropical conservatories with considerable success. In British Columbia, for instance, *Planococcus citri* was controlled with *Cryptolaemus montrouzieri* and *Leptomastix dactylopii*, *Tetranychus urticae* with *P. persimilis*, *Panonychus ulmi* with *Metaseiulus occidentalis* and *Amblyseius californicus*, and *Trialeurodes vaporariorum* with *Encarsia formosa*. Experiments with similar pests are being conducted in other public conservatories, butterfly farms and botanic gardens.

The cold temperatures often experienced in Northern European plantings are likely to limit the use of exotic parasites to the summer.

In summary, control systems are based round careful inspection of plant material before purchase and the use of natural enemies, while chemical treatments are harmonized as best as possible with the natural enemies present. The diversity of plants and these different situations involved will mean that each must be dealt with individually.

6.5 PEPPERS
N.W. Hussey

In the Netherlands, where more than 30% of the 180 ha of the crop production are protected by biological control of pests, a simple but effective integrated programme has been developed.

The control of spider mites, of course, depends on *Phytoseiulus* while thrips are contained by *Amblyseius cucumeris*. It is most important to maintain thrips at insignificant levels because even small numbers can cause serious damage by feeding below the calyx of the fruit where a rot is initiated. Bearing in mind the cheap methods for producing the predator, it should be economic to make repeated introductions of the predator as an 'insurance' using bran (Spidex®) as a dispersant (Figure 39).

Myzus persicae is a serious pest unless controlled by pirimicarb, though *Aphidius matricariae*, a native parasite, usually becomes established in the absence of broad-spectrum insecticides and contributes to the control.

Tarsonemid mites may become a problem in damp conditions and at the first sign of damage – puckering of the leaves – dicofol should be applied.

Bacillus thuringiensis or dimilin provide effective control of noctuid caterpillars without affecting predatory mites.

The principal diseases, *Botrytis* and *Rhizoctonia*, are selectively controlled with vinclozolin.

Such a harmonization of chemical and biological control measures is a model which should be sought on other minor crops which are attacked by a relatively narrow range of pests and diseases.

7. IMPLICATIONS FOR BIOLOGICAL CONTROL ON INTENSIVELY GROWN OUTDOOR CROPS

7.1 STRAWBERRIES

N.E.A. Scopes

Strawberries are widely grown throughout the world and, for early production in temperate climates, they are frequently cultured under polythene. This situation encourages red spider mite (*Tetranychus urticae*) to become an even more troublesome pest than it is on unprotected field-grown crops.

It has been established, both in California and in the UK, that biological control of this pest outdoors is feasible, though the numbers of predators required made it too costly. However, a recent series of trials by the GCRI and ADAS resulted in a more rational approach to controlling this pest. The principle underlying the trials was similar to that used in the control of red spider mites on cucumber crops in greenhouses. Since spider mites overwinter after the adult has entered diapause, populations must be controlled in the late summer before the onset of this hibernation. Additionally, suitable hibernation sites must be eliminated as soon as possible to reduce the potential for carry-over to the next season.

Red spider mites hibernate in protected situations, such as straw, debris and the crowns of plants. They emerge in the spring and begin feeding and multiplying on the older leaves. Considerable reductions in pest numbers may be achieved by cleaning up the land to destroy debris. Fumigating soil with methyl bromide is possible on small areas but much of its effectiveness may be lost if newly purchased runners are already infested with mites.

Once mites are established on plants, it is virtually impossible to eliminate them with chemicals and so extra care has to be taken to purchase 'pest-free' runners. It has proved possible to treat cold-stored runners with methyl bromide and so eliminate red spider mites (12–$44\ g/m^3$ for 2 hours at room temperature), but careful dosing is required so as not to harm the dormant plants.

Current commercial methods of using *Phytoseiulus persimilis* are based on the results of the collaborative trials and have proved successful both on crops grown in polythene tunnels and in the open (Port & Scopes, 1981).

The first introduction of predators (1/plant) should be made in August or early September, shortly after planting, and, with the warm temperatures that usually prevail, most of the red spider mites should be eliminated before they are stimulated to hibernate. Predators may sometimes naturally colonize the planting if they have been used regularly before, so allowing some economy in the autumn introduction.

A second introduction of predators (1/plant) should be made in the following spring, usually in mid-March, when the first mites emerge from hibernation or when the likelihood of frosts has disappeared. It is particularly important to establish predators early in the season so that they can multiply to cope with any unforeseen build-up of mites.

Spring treatments with predators are unlikely to be commercially successful if the autumn treatment has been omitted as the large overwintering mite population leads to rapid increase of the mites on the few early-developing leaves so that damage would ensue before predators gained control.

This system, in common with all biological pest management programmes, requires monitoring, and introductions in the second and subsequent seasons are likely to require even fewer predators. Success depends on the grower carefully checking his crop and spotting the first signs of mite activity on the crop.

Chemical treatments against the other serious pests and diseases, e.g. aphids and *Botrytis*, must be made with compatible chemicals. Experiments are in progress to evaluate *Amblyseius mackenziei* as a potential control of tarsonemid mites, which are becoming more serious on strawberries.

REFERENCES

Port, C.M. & Scopes, N.E.A. (1981) Biological control by predatory mites (*Phytoseiulus persimilis* Athias-Henriot) of red spider (*Tetranychus urticae* Koch) infesting strawberries grown in 'walk in' plastic tunnels. *Pl. Pathol.* **30**: 95–99.

7.2 HOPS (*HUMULUS LUPULUS* L.)
C.A.M. Campbell

The damson-hop aphid (*Phorodon humuli*) and the spider mite (*Tetranychus urticae*) are hop pests of almost world-wide importance. *P. humuli* is generally of more concern to growers in Europe whereas, in the USA, *T. urticae* is always a more serious problem. *P. humuli* has not been reported from Australasia. Other arthropod pests that are substantially less important include the weevil, *Otiorhynchus ligustici*, the flea beetle (*Psylliodes attenuata*) and caterpillars of the moths (*Pyrausta nubilalis*, *Hepialus sylvina*, *Pseudosciaphila branderiana* and *Cnephasia wahlbomiana*). Common root-feeding nematodes include *Xiphinema diversicaudatum* which, as a vector of arabis mosaic virus, is always a threat to hop growing, and the cyst nematode (*Heterodera humuli*).

Both *Phorodon humuli* and *Tetranychus urticae* are capable of destroying crops and this has resulted in a long history of chemical control in which the contribution of natural enemies has been minimal. Frequent applications of pesticides have provided unrelenting pressure to select individuals that are most resistant to those pesticides. Both pests now show widespread resistance to compounds from most of the major chemical groups (Hrdy & Hrdlickova, 1981). Consequently there is a precarious reliance on very few pesticides for chemical control. The development of resistance also stimulated interest in alternative strategies to obtain commercially acceptable levels of control.

Work in the USA (Peters & Berry, 1980) has shown that some hop varieties exhibit potentially useful resistance to *T. urticae*. Of potentially wider interest, also from research in the USA (Pruszynski & Cone, 1972), is the possibility of supplementing the numbers of the phytoseiid, *Typhlodromus occidentalis*, a predator of *Tetranychus urticae*, by controlled releases. *Typhlodromus occidentalis* has been widely studied for the biological control of spider mites on a range of outdoor horticultural crops. In New Zealand, it controlled infestations of *Tetranychus urticae* on strawberries and apples in the season of release (Penman *et al.*, 1979). As some species of phytoseiids are now being bred in commercial quantities for the control of *T. urticae* in greenhouses, the way seems clear for similar development on hops.

Fairly accurate predictions are now possible for the commencement and duration of the migration of winged aphids from *Prunus* spp. to hops (Thomas *et al.*, 1983), the sole secondary host of *Phorodon humuli*. Typically migration extends from late May until mid-July. On hops the aphid does not produce winged virginoparae capable of infesting other hops. Therefore, the predictions give valuable warning of the period that hops are at risk from migrant aphids, and allow more timely counter-measures. Aphid predators can establish themselves on hops only after the aphids have arrived and begun to reproduce. In the UK, experience has shown that, unless a pesticide is used to limit the early season build-up of aphids, then the inevitable delay before the arrival of predators almost always leads to overpopulation by the pest, closely followed by defoliation of the plants and the loss of any harvestable crop (Campbell, 1977). However, it was found that, if the initial build-up of aphids was contained by using a relatively ineffective insecticide, then sufficient aphids survived to attract predators and they regulated numbers at sub-economically important levels for the rest of the season. These findings have formed a framework for the development of integrated control of the aphid.

Current research is on two fronts. Firstly, suitable pesticides are being screened for use early in the aphid's migration period and other selective pesticides are being sought for those sites and occasions when the natural influx of predators is inadequate to prevent damage (Cranham *et al.*, 1982). Secondly, further possible biological constraints to *Phorodon humuli* are being evaluated. In England, earwigs (*Forficula auricularia* L.) and anthocorid bugs (*Anthocoris* spp.) are the most important natural enemies in hop gardens (Campbell, 1978), whereas in Central and Eastern Europe, coccinellids and predacious Diptera may be more important (Hrdy & Hrdlickova, 1981). The lack of synchronization between the arrival of aphids and these predators on hops, and the uncertainty as to whether sufficient numbers of predators will be attracted into the crop, could be avoided by controlled introductions of selected natural enemies. Parasitoids, predatory arthropods and fungal pathogens are all being considered for this role. The fungal pathogen, *Verticillium lecanii*, which is being tested for aphid control in greenhouses was very effective against *Phorodon humuli* in greenhouse tests, but proved ineffective in the field (Cranham *et al.*, 1982).

As yet, no outstandingly high levels of resistance to aphids have been found among hop plants in germ plasm collections, although varieties do vary both in their suitability for aphid multiplication (Campbell, 1983) and in their attractiveness to the winged migrants. Further investigations aim to reduce the numbers of winged aphids that settle on tops by using sprays of an aphid alarm pheromone, or by exploiting the visual non-preference of migrant aphids for landing on certain colour forms of their host plants.

There have been no reports of experiments to exploit biological controls for the minor arthropod pests of hops. Fungi are important natural control agents for hop cyst nematodes (Mikhajljukov, 1976), and fungal agents are now being considered for the control of both *Heterodera humuli* and *Xiphinema diversicaudatum* (McNamara, personal communication, 1983).

REFERENCES

Campbell, C.A.M. (1977) Distribution of damson-hop aphid (*Phorodon humuli*) migrants on hops in relation to hop variety and wind shelter. *Ann. Appl. Biol.* **87**: 315–325.

Campbell, C.A.M. (1978) Regulation of the damson-hop aphid (*Phorodon humuli* (Schrank)) on hops (*Humulus lupulus* L.) by predators. *J. Hort. Sc.* **53**: 235–242.

Campbell, C.A.M. (1983) Antibiosis in hop (*Humulus lupulus*) to the damson-hop aphid, *Phorodon humuli. Entomologia Exp. Appl.* **33** 57–62.

Cranham, J.E., Souter, E.F., Tardivel, G.M. & Firth, S.I. (1982) Damson-hop aphid, *Phorodon humuli* (Schr.). *Rep. E. Malling Res. Stn* **1981**: 105–107.

Hrdy, I. & Hrdlickova, H. (1981) Integrated pest and disease control in hops. *Bull. IOB/WPRS Panel on Integrated Pest and Disease Control in Hops, Liblice* **4** (3): 179 pp.

Mikhajljukov, V.S. (1976) Mycoses of cysts of *Heterodera humuli* Filipjev, 1934. *Mikrobiol. Zh.* **38**: 173–175.

Penman, D.R., Wearing, C.H., Collyer, E. & Thomas, W.P. (1979) The role of insecticide-resistant phytoseiids in integrated mite control in New Zealand. pp. 59–69. In *Recent Advances in Acarology* (Rodriguez, J.G., ed.) Volume I. New York, Academic Press. 631 pp.

Peters, K.M. & Berry, R.E. (1980) Resistance of hop varieties to two spotted spider mite. *J. Econ. Ent.* **73**: 232–234.

Pruszynski, S. & Cone, W.W. (1972) Relationships between *Phytoseiulus persimilis* and other enemies of the two spotted spider mite on hops. *Envir. Ent.* **1**: 431–433.

Thomas, G.G., Goldwin, G.K. & Tatchell, G.M. (1983) Associations between weather factors and the spring migration of the damson-hop aphid, *Phorodon humuli. Ann. Appl. Biol.* **102**: 7–17.

7.3 BLACKCURRANTS
M. Easterbrook & M. Lyth

In the UK, the main developments in non-chemical methods of control for pests of blackcurrant have been in the field of breeding for resistance. For many years, the most important pest of blackcurrants has been the blackcurrant gall mite (*Cecidophyopsis ribis*), which causes the characteristic 'big buds'. The mites themselves can cause some loss of yield, but they are of greater importance as vectors of reversion virus, which can quickly kill the plant. At present only one chemical, endosulfan, is approved for use against the mite in the UK, so alternative control methods are particularly desirable. Because of its long history of freedom from galled buds and reversion virus, the gooseberry (*Ribes grossularia*) was selected as the main donor of resistance to the mite. A dominant gene *Ce*, controlling resistance to the mite, was transferred from gooseberry to blackcurrant (Knight *et al.*, 1974). Since then backcrosses have been made to improve agronomic characters and confer disease resistance (Keep *et al.*, 1982) and selections suitable for commercial release should be available shortly. There are also several known sources of host resistance to aphid pests of *Ribes*. Accessions of *R. sanguineum* and *R. glutinosum*, carry high levels of recessive resistance to the currant-sowthistle aphid (*Hyperomyzus lactucae*) and a number of other *Ribes* pests. Clearly breeding for resistance to pests or to the viruses themselves, e.g. to reversion (Keep *et al.*, 1982), offers a promising alternative to chemical control.

At present, biological control is not used against any pests of blackcurrant in the UK. However, some possibilities exist for future developments, in particular the use of entomophilic nematodes. One species, *Neoaplectana bibionis* Bovien, has been used in Tasmania to disinfect blackcurrant cuttings of the larvae of *Synanthedon tipuliformis*, the currant clearwing (borer) moth (Bedding & Miller, 1981), and it may prove possible to use a spray containing nematodes on established plantations. The use of nematodes also offers a prospect for control of the vine weevil (*Otiorhynchus sulcatus*), which is becoming an important pest of blackcurrants in the UK, and which is difficult to control by chemical means. The fungus, *Verticillium lecanii*, which has been used for biological control in greenhouses, has been found in field-collected samples of blackcurrant gall mite and, in laboratory tests, mites were killed by an application of spores of the fungus (Kanagaratnam *et al.*, 1981). It is possible, therefore, that biological control could be achieved in blackcurrant plantations by artificial application of spores during the migration period of the mites.

REFERENCES

Bedding, R.A. & Miller, L.A. (1981) Disinfecting blackcurrant cuttings of *Synanthedon tipuliformis*, using the insect parasitic nematode, *Neoaplectana bibionis*. *Envir. Ent.* **10**: 449–453.

Kanagaratnam, P., Hall, R.A. & Burges, H.D. (1981) Effect of fungi on the blackcurrant gall mite, *Cecidophyopsis ribis*. *Pl. Pathol.* **30**: 117–118.

Keep, E., Knight, V.H. & Parker, J.H. (1982) Progress in the integration of characters in gall mite (*Cecidophyopsis ribis*) – resistant blackcurrants. *J. Hort. Sci.* **57**: 189–196.

Knight, R.L., Keep, E., Briggs, J.B. & Parker, J.H. (1974) Transference of resistance to blackcurrant gall mite, *Cecidophyopsis ribis*, from gooseberry to blackcurrant. *Ann. Appl. Biol.* **76**: 123–130.

8 MASS PRODUCTION OF NATURAL ENEMIES
N.E.A. Scopes & R. Pickford

In the early days of the development of biological control, the GCRI published a booklet designed to advise those growers who wished to rear their own natural enemies. With slight modifications, this Section reproduces the content of that booklet (now out of print) but also includes notes on the methods used by one of the major rearing companies.

The successful propagation of natural enemies demands keen observation, patience, practice, attention to detail, foresight and continual care, though it is *not* necessary to have entomological expertise. The facilities described exceed the minimum requirements, but they are essential if continuous production is to be guaranteed.

8a **WHEN IS 'REAR YOUR OWN' A JUSTIFIED POLICY?** Home production provides an insurance against sudden pest outbreaks; these can be treated immediately without worry of delays in delivery from commercial suppliers. Furthermore, since biological control programmes must be based on 'blueprint' growing, the actual environmental conditions (i.e. amount of sunshine) may upset the intended synchrony between pest and natural enemy and so advance orders placed with commercial suppliers may not deal with the problem so effectively as your own supply.

The convenience of having biological agents continually available must be contrasted with the likely cost of control by chemicals or with the purchase of natural enemies from commercial suppliers. On these terms home production from a large breeding unit will not appear attractive for small acreages, though small groups of growers should carefully consider the possibility of collective action. The capacity of the elaborate, large rearing system described is considerable – a minimum of 2 ha/week for predators and 1 ha/week for *Encarsia* parasites – and so a saleable surplus could provide additional revenue for part of the year.

If the decision is made on cost grounds alone, when compared with chemical control, a large unit would be justified only if the holding has more than 1 ha of glass and this area would increase to 2 ha if comparison is made with the cost of purchased parasites and predators.

The less sophisticated systems which will be described require a clean plant house, red spider and predator house (separating cultures by water barriers) and two *Encarsia* houses. They would be financially attractive to growers with *much* smaller acreages and heating costs can, of course, be reduced by lowering the greenhouse temperature, although this will reduce output since it will slow down the rate of development of the natural enemies.

General requirements Rearing techniques comprise three distinct processes:

1. Production of clean (uninfested) plant material.
2. Production of the host (pest) organism on these clean plants.
3. Production of the natural enemy.

All three processes must be rigidly isolated although each stage is dependent on the previous one: for instance, any irregularities in clean plant production could be reflected in the output of natural enemies some weeks later. Familiarization with the rearing of each natural enemy species will help to ensure successful control on the crop plants for, although there are 'rule of thumb' guides, it is essential to 'get the feel' of biological control techniques if problems are to be anticipated.

GREENHOUSES The use of small individual houses is preferred to larger compartmented structures. They permit safe fumigation and provide maximum versatility. The houses should be spaced at least 4 m apart and connected by smooth paths to facilitate the use of trolleys for moving plants. The greater the space between the greenhouses, the less the chance of contamination by predators or parasites. Without duplication of rearing facilities, premature contamination of clean pest cultures by parasites or predators could take at least 3 months to remedy. It is vital to programme the work so as to avoid insects being carried between houses on clothing and equipment.

LIGHTING Supplementary lighting is essential for the maintenance of red spider mite stocks and an intensity of 5400 lux, for 16 hours per day, is necessary both to prevent hibernation and to maintain reproduction during short winter days. A bank of fluorescent tube lighting (350–400 watt) is sufficient to illuminate 3 m^2 of bench. Lighting is unnecessary for predator cultures though supplementary lighting for 16 hours a day benefits whitefly in the winter. A 300 watt lamp is sufficient for 7 m^2 of floor area, though sealed lighting units should be used to prevent whiteflies exuding honeydew over electrical contacts. While lights are not essential for the production of most clean plants, the growth of tobacco (for *Encarsia* production) should be improved by supplementing the poor light on dull winter days.

USE OF INSECTICIDES Insecticides are invariably more harmful to natural enemies than to pests; they must, therefore, always be used with extreme care. Any new pesticide or method of application must be thoroughly tested for undesirable side effects.

Larvae of sciarid flies are often troublesome in the culture of pot plants so the potting compost must be drenched with solutions containing 5 cc diazinon 20% liquid in 9 l water or 1 g diflubenzuron in 1 l water before sowing or pricking out seedlings. Protection should last for up to 6 weeks.

Aphids, especially on tobacco and broad bean plants, may be safely controlled with a soil drench of pirimicarb (50 ml/25 cm pot).

8b PRODUCTION OF RED SPIDER MITE PREDATORS (*PHYTOSEIULUS PERSIMILIS*) Three greenhouses or compartments are required for:

1. Clean plant production.
2. Red spider mite production.
3. Predator production.

Both broad beans and French beans are used, the former because they produce more spider mites per unit area of bench space, and French beans because their large flat leaves are better suited for the production and distribution of predators.

CLEAN PLANT PRODUCTION Twenty-eight broad beans (*Vicia faba*) are sown in 4 rows in seed boxes. When the first shoots appear, usually after less than a week, the boxes are transferred to a second greenhouse for infestation with red spider mites.

Dwarf French beans (*Phaseolus vulgaris*) are sown (4 seeds/25 cm diameter pot) and grown for 4–5 weeks until each plant has produced 4–5 true leaves. The growing points should then be removed to expand the total leaf area and the plants staked before infestation with spider mites.

RED SPIDER MITE PRODUCTION When the first green shoots appear each tray of broad beans is infested with spider mites using three heavily infested stems from older bean plants (Figure 35). The beans are kept upright and firmly supported by 1 m bamboo canes placed in each corner of the seed tray and tied together both at their apex and 15 cm below (Figure 36). After 3–4 weeks when the first spider mites aggregate on the apical leaves, the foliage is removed and placed on 5 or 6 pots of French beans. As the leaves dry the mites migrate onto the new host. Once infested, these plants are ready for transfer to a third greenhouse for infestation with predators.

Dichlorvos (DDVP) slow-release strips (one strip to 30–50 m^3) may be used to keep red spider mite cultures free from premature contamination by predators – 2 treatments at 4-day intervals are usually required. Overnight fumigation kills active stages of the predator though pro-

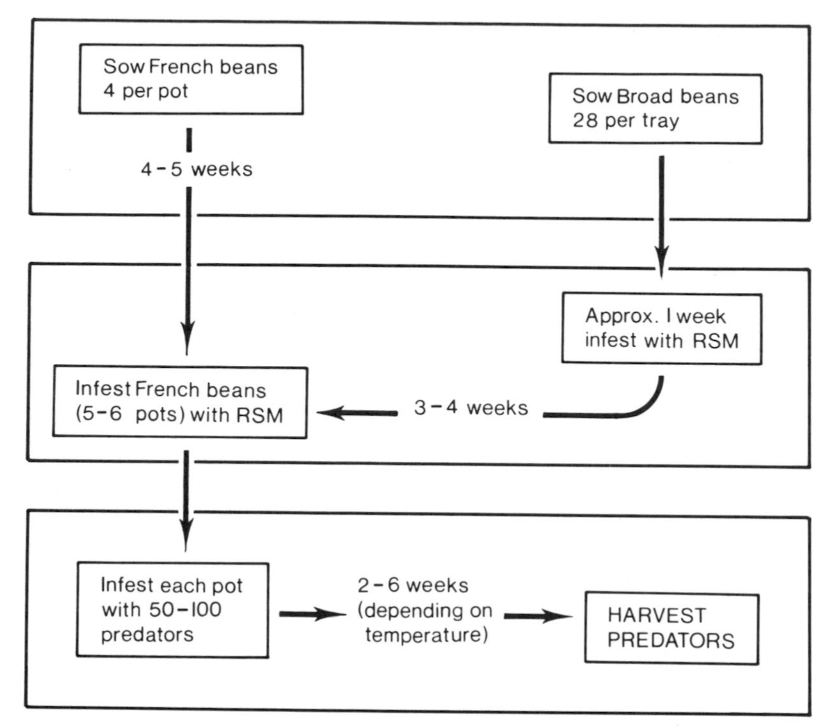

FIGURE 35 Procedure for mass-producing the predator, *Phytoseiulus persimilis.*

longed treatments are needed to kill eggs. These treatments also affect red spider mites and should therefore be used sparingly.

PREDATOR PRODUCTION Each pot of French beans is infested with 50–100 predators on a piece of infested leaf. Harvesting (1500–4000 predators/pot) is most efficiently achieved when the red spider mite population is almost eliminated. The time from infestation to harvest averages 2–2½ weeks at 22°C though the process may be slowed down at lower temperatures, 3–3½ weeks at 18°C or 5–6 weeks at 12°C.

The cultures should be isolated from each other and harvested at the correct time. If the predators are not required the plants must be destroyed. This prevents vast numbers of predators from migrating in search of food and, within a few days, devouring mites on newly infested plants thereby disrupting the production programme.

At the GCRI, some cultures are maintained over trays of water, which act as a barrier to predator migration. Care must, however, be taken to prevent foliage touching the sides of the greenhouse or that of other plants. In this way, cultures of different ages are always available. At harvesting, predator-infested leaves are put into polythene bags for distribution. Storage of predators in these bags at 8–10°C is possible provided that condensation is prevented by the insertion of blotting paper or newspaper and adequate food is available in the form of mite-infested French bean leaves.

Assuming that each pot of mature French bean plants occupies approximately 0.1 m² of bench space, the use of 3 × 2.5 m glasshouses for both spider mite and predator production should enable at least 15 pots to be harvested weekly, so providing 22 500–60 000 predators, sufficient for at least 2 ha. On one large rearing unit, 7 houses are used in succession so that the whole process from plant propagation to harvest is completed in each house. Another innovation is to use the runner bean cv. 'Scarlet Emperor' so that the whole internal volume of the house is utilized for production. In the propagating houses, 8 beans are germinated in each 25 cm plastic pot beneath the benches and then spend 2 weeks on the bench before introduction to a production house where, after 1 week, they are infested with spider mites which are allowed to develop for 2 weeks. In week 4, the predator is introduced and harvested during week 6. The cycle is then repeated so that, in week 7, a house is cleaned before receiving a fresh intake of young beans.

Lighting uses are 400 watt sodium bulbs to every 11 m². Pots of soil-less compost are watered by capillary matting on a concrete floor. Each house is heated by automatically controlled hot water system and ventilated automatically by extraction fan. Propagating houses are heated to 15°C while ventilating begins at 23°C. Those houses concerned with predator production are ventilated at . 25°C. Predators are eliminated from new production units (week 2) by putting a few drops of dichlorvos in a tin tray on the heating pipes for 2 days.

FIGURE 36 Mass production of red spider mite on broad beans. N.B. Canes to support plants and collect mites.

8c PRODUCTION OF WHITEFLY PARASITES (*ENCARSIA FORMOSA*)

Three methods of production are described – one is sophisticated and precise and is intended for the larger growers, and the others, while less demanding, are perhaps more suitable for small growers. A comparison of the requirements for two different scales of production is set out in Table 9.

TABLE 9 Comparison of production systems. Figures given are a minimum and experience would increase production dramatically

System	Production per week	Area required (m^2)	Greenhouse requirement	Temp. (°C)	Lights	Bugs/ m^2/wk
Phytoseiulus						
Greenhouses	20 000+	25	3 houses (4 × 3 m)	21 ± 5	Yes (RSM)	
Trays	2500+	2.7	Spare space	21 ± 5	Yes (RSM)	
Commercial production	300 000	120	6 (4 × 7 m)	15	Yes	2500
Encarsia						
Greenhouses (for 1 ha+)	140 000	120	5 houses	24	Yes (Whitefly)	
Cheshunt method (for 0.1/1 ha)	10–20 000	0.3 m^2/ plant	7 × 4 m	21 ± 5	No	
Cages (for 0.1 ha)	5000	3	Spare space	21 ± 5	Yes	
Commercial production	1 700 000	130	7 (2 × 3 × 6 m & 0.3 × 3 × 4 m)	15–20	Yes	13 000

Method 1 (for large growers with more than 1 ha to treat) Whitefly, and subsequently the parasites, are produced on tobacco plants (*Nicotiana tabacum* cv. 'Brazilian') in 5 separate greenhouses maintained at 22°C. One house, required to produce the clean plants, could also produce French and broad bean plants for predator production. The other 4 houses are used for different stages of the whitefly and parasite production process. At the GCRI we also use a covered hand truck which acts both as a fumigation cabinet and plant transporter. Cool storage facilities (12–15°C) are useful though not essential. Plastic pots must not be used to culture plants for *Encarsia* production because they absorb dichlorvos which may subsequently be released in quantities sufficient to kill adult parasites.

CLEAN PLANT PRODUCTION Tobacco seeds are sown and grown in 18 cm diameter clay pots at 22°C until they produce 10–12 leaves (6–9 weeks). To withstand severe whitefly attack tobacco plants must be regularly watered and fed. To stimulate the growth of young seedlings in winter supplementary lighting is beneficial, though not essential.

Encarsia PRODUCTION Six tobacco plants (● ● in Figure 37) are infested by the vast number of adult whiteflies maintained in the whitefly-production house. The plants should be gently agitated 4 or 5 times during the period of exposure to ensure an even distribution of adults, and hence eggs, on the leaves. During the summer, 8 hours infestation

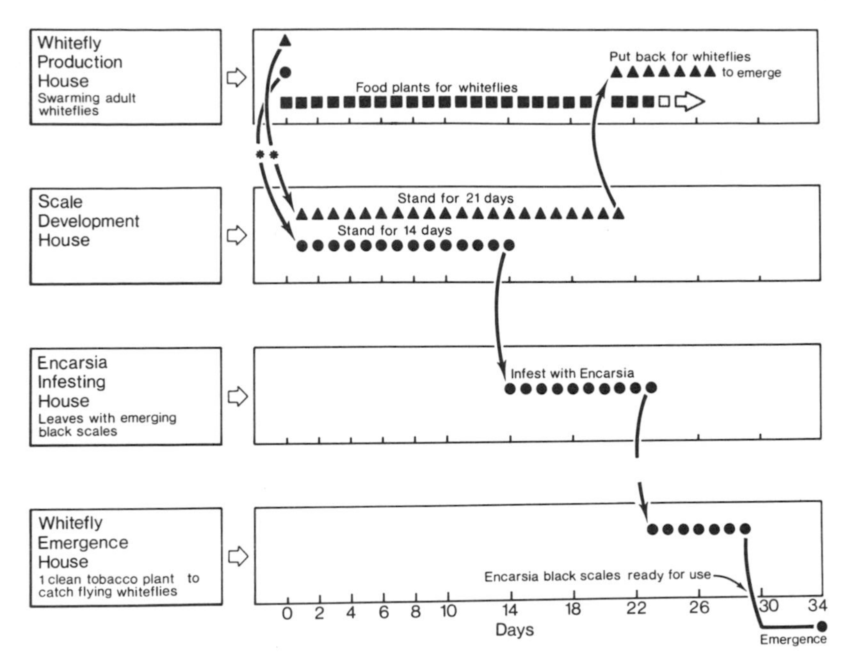

FIGURE 37 Procedure for mass-producing the whitefly parasite, *Encarsia formosa.*

should be sufficient to ensure adequate numbers of eggs though this period may have to be extended to 24–48 hours during the winter, especially in dull weather. No more than 40 eggs/cm² are required on each leaf if excessive honeydew production by the scales is to be avoided. The plants are then shaken to remove most whiteflies and fumigated with dichlorvos slow release strip (1/56 m³) to kill the remainder. Though this treatment should kill adults within 4 hours, extended fumigation, even up to 4 days, does not appear to harm whitefly eggs. Prior to standing treated plants in the scale-development house their growing points should be removed to stimulate leaf expansion. The scales develop to a stage suitable for parasitism by *Encarsia* after 2 weeks (Figure 38). (Visually this occurs when the 'scale' ceases to be a flat disc and develops elevated sides.) The plants are then transferred to the *Encarsia* infesting house and infested with parasites by hanging 2–3 leaves, bearing 4000–5000 black scales among them. To ensure a high level of parasitism (95%+) between 1 and 4 wasps must be available to search each 6.5 cm² of leaf surface. After about 9 days, when the first black scales have formed the plants are removed and fumigated with dichlorvos for 1 hour to reduce the risk of contaminating the rearing unit with parasites. The plants are then placed in the whitefly emergence house where, 17–29 days after original infestation, any unparasitized whiteflies emerge. The plants are again fumigated with dichlorvos for 2–4 hours to kill these adults, so leaving only black, parasitized scales. (This fumigation can be omitted if complete parasitism has occurred.) A clean tobacco plant, changed fortnightly, is kept in the whitefly emergence house to catch any stray adults.

The plants, now bearing only parasite pupae (black scales), are now ready for use. Each should have 5–8 leaves populated by up to 15 000 black scales. Parasites should emerge 10–12 days after the scales turn black, though cool storage will prolong this period. At 12°C newly formed, black scales can be kept for about 4 weeks, while at 16°C 2–3 weeks successful storage may be expected.

MAINTENANCE OF WHITEFLY STOCKS The production of large numbers of whitefly eggs depends on maintaining large populations of adults. The system, operating every week, consists of two processes: firstly, the production of whiteflies to maintain stocks and secondly provision of food to keep them alive.

Three plants (▲▲ in Figure 37), similarly infested with whiteflies, are fumigated and stood in the scale development house for 3 weeks. They are then transferred back to the whitefly production house where the adults emerge 6 days later (approximately 27 days after egg laying).

Three more clean tobacco plants (■■ in Figure 37) are put into the whitefly production greenhouse to provide host plants for the flies. These normally remain suitable as a food source for 3 weeks and after 4 weeks their leaves are hung in the glasshouse for a further week to allow any pupating whiteflies to emerge.

If this large scale production method is operated in four 4×3 m greenhouses together with a clean plant greenhouse (at least 5×3 m having an input of 6 plants per week (for *Encarsia*), about 100 000 black parasite scales (≡supply for 5–6 ha) should be produced each week. More plants can be used so long as sufficient adult whiteflies are available to infest them, and no doubt the programme could be repeated twice weekly to double production, though extra space would then be needed, particularly in the clean plant and whitefly infesting greenhouse.

Method 2 (for small growers up to 1 ha) An alternative method is that used for many years at the former Cheshunt Experimental Station. According to the number of parasites required, the rearing unit should consist of one or two small greenhouses or compartments, plus a small area for propagating clean plants. The parasites are reared on tomato plants, while tobacco plants (cv. 'White Burley') are used to produce the large numbers of whiteflies necessary to maintain adequate parasitism. This method is less critical and less demanding than the one already described and hence more suitable for the small grower. One house (about 7 × 4 m) should provide about 15 000–20 000 parasites weekly (400 per plant), sufficient to treat 1 ha of greenhouses over a period of 1 month. Where two or more rearing houses are used they should preferably be operated in sequence with an interval of about 6 weeks between planting and stocking each. This helps to provide an uninterrupted flow of parasites.

After propagating, tomato plants in 4 cm whalehide pots are set out in the rearing house(s) at normal commercial spacing in double rows. When about 30 cm tall, the plants are infested with whiteflies by introducing heavily infested pots of tobacco amongst them at a ratio of about 1 tobacco to 10 tomato plants. Two weeks later, parasites are introduced by hanging up leaves bearing black, parasitized scales. Once the interaction is under way it should be self-perpetuating. As the plants grow, so the whiteflies and parasites move up the plants, providing a succession of leaves populated by black scales. These leaves are removed for use on the commercial crop. However, to ensure that sufficient parasites are maintained in the rearing houses some black scales should always be left until the adults have emerged.

The temperature in the rearing houses should be maintained at 22°C (minimum of 18°C).

If parasite production is required throughout the year, the rearing houses should be replanted and restocked about every 4 months.

Maintenance of the correct balance between the numbers of whiteflies and parasites in the rearing unit can be achieved only with experience. Excessive numbers of whiteflies can usually be removed by reducing the number of tobacco plants or by fumigating with a non-persistent insecticide such as propoxur smokes or dichlorvos strips. Spraying the tops of the tomato plants with pyrethrin insecticides also effectively reduces numbers of adult whiteflies.

Method 3 (for small growers up to 0.2 ha) Parasites can be reared on plants enclosed in separate wood-framed, gauze-covered (50 mesh) cages 0.6 × 0.6 m in area and 1.6 m high kept in a heated greenhouse, provided great care is taken to avoid escapes to the crop. Each plant is infested at the 2–3 leaf stage with about 300 adult whiteflies. Twelve (22°C) to 20 (18°C) days later, depending on temperature, pieces of leaf bearing about 600 black, parasitized scales are introduced into each cage. Four to 5 weeks later each plant should begin to produce parasitized scales and about 5000 should be harvested over 3 weeks. To ensure continuity of parasite production, plants should be set up in sequence at weekly intervals, thus requiring 6–7 cages. Care must be taken not to overwater the plants or allow them to dry out and the cages should be shaded during spells of hot sunny weather.

FIGURE 38 Mass production of parasites. Oviposition by whiteflies (*left*). Whitefly scales parasitized by *Encarsia* (*right*).

Method 4 (commercial production) Three houses are used for propagating tobaccos in 25 cm pots which, in winter, takes 13 weeks. The plants are then introduced to the whitefly emergence house which is heated to 15°C, although, for a few hours after introducing the plants, the temperature is raised to 25°C. Ventilation is started if the temperature reaches 30°C. The lighting in this house is in the form of fluorescent tubes rather than the sodium lamps used in the rest to avoid death of adults by attraction to the lights. After a week, adults are blown off the plants with a vacuum cleaner and then moved to one of five other houses (operated at 20°C) for a further week to allow the scales to develop to instar 2–3 and so be attractive to parasites. Premature invasion by the latter is prevented by a 2-day fumigation with dichlorvos, as for predators.

Parasites are introduced on leaves hung on strings above heating pipes along the sides of the house. The winter temperature of 20°C is raised to 25°C for the first 3 days, with ventilation at 30°C. Throughout the 4 weeks of parasite activity, the plants are washed down daily with a hosepipe to remove honey dew.

FIGURE 39 Rearing room at GCRI. Each cage is ventilated by cross-flow of temperature-controlled air drawn from the room and is lit independently by different light banks.

After 4 weeks the black parasitized scales are ready for harvest. The lower leaves are removed in this process but the youngest leaves are left, still infested with adult whiteflies, and provide fresh plants for the whitefly emergence house.

An additional house is used to preserve the whitefly stock with tobaccos which spend 3 weeks alternately in the emergence and production houses. Where pure nuclear stocks of natural enemies have to be maintained, the simple, yet relatively inexpensive, system at the GCRI (Figure 39) could be utilized.

8d **PRODUCTION OF THE APHID PARASITE (*APHIDIUS MATRICARIAE*)** *Myzus persicae*, the peach-potato aphid, is a common pest on chrysanthemums, ornamental pot plants and peppers, which can be effectively controlled with small numbers of the parasitic wasp (*Aphidius matricariae*).

CLEAN PLANT PRODUCTION Chinese cabbage (cv. 'Pte Sai') is sown and grown singly in pots to produce robust plants which are ready for infestation with aphids after 4–7 weeks.

APHID PRODUCTION Two gauze-covered cages approximately 1×0.75 m and about 0.5 m high should be accommodated in a glasshouse. Seven plants should be caged every week and each infested with 10–20 aphids by introducing a piece of aphid-infested leaf. At 22–25°C about 1000 aphids will be produced after 2 weeks. One plant is kept to maintain the aphid stocks, while the other 6 are used to produce parasites.

Aphidius PRODUCTION Three similar gauze cages are required to produce the parasites. Every week, 6 aphid-infested Chinese cabbage plants are introduced into a cage together with about 30 mummies (parasite pupae). The adult parasites emerge, sting and lay eggs in the aphids; $1\frac{1}{2}$–$2\frac{1}{2}$ weeks later (22–25°C) mummies develop and the plants are harvested by removing the leaves, washing off live aphids and leaving the mummies to dry. Leaves bearing mummies should be hung in commercial greenhouses to allow adult parasites to emerge.

The parasite production cages should be isolated from those used for rearing aphids, preferably in a separate greenhouse. Each cage of Chinese cabbages should produce 3000–6000 parasites, sufficient to treat 0.1 ha.

It is difficult to judge when a particular rearing method will be adopted for large-scale production and so, in the following cases, proposed

FIGURE 40 Packs used to distribute *Encarsia formosa* in the Netherlands. The parasitized scales are glued to cards and provided with a source of honey (*left*). The predator *Phytoseiulus* is distributed with its prey in bran (*right*).

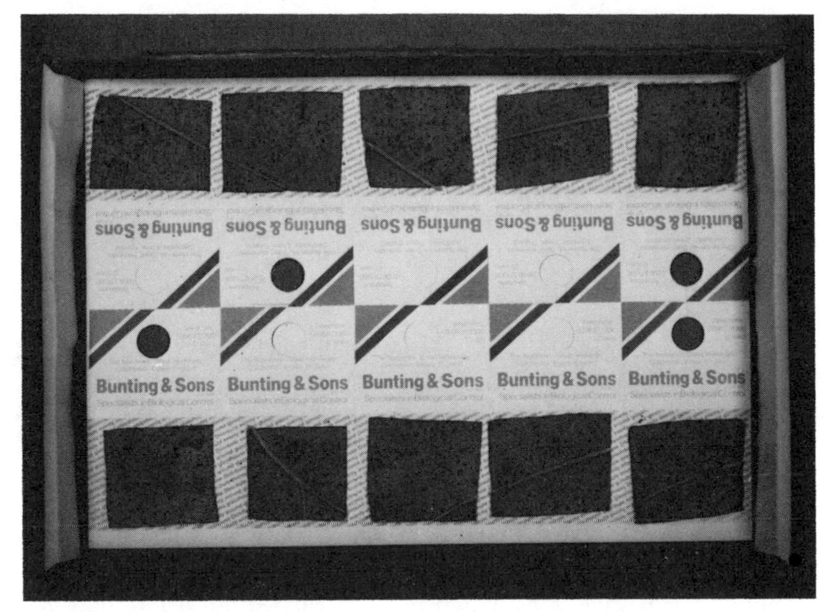

FIGURE 41 Distribution of *Encarsia* in the UK. The parasitized scales on the host leaf are stuck on cards.

rearing techniques have been outlined in the text where the potential use of the natural enemy concerned is discussed:

1. Rearing of the mealybug predator (*Cryptolaemus montrouzeri*) – Section 2.9.
2. Rearing of the aphid-predator (*Aphidoletes aphidimyza*) – Section 2.8.
3. A method for providing supplies of *Phytoseiulus* at short notice – Section 4.3.

Natural enemies are sold and distributed by different methods (Figures 40 & 41) by different companies but all now provide a reliable service to the customer.

REFERENCES

De Bach, P., edit. (1964) *Biological Control of Insect Pests and Weeds* Reinhold Publishing Company, New York. 844 pp.

King, E.G. & Leppla, N.C., eds (1984) *Advances and Challenges in Insect Rearing*, USDA–ARS. 306 pp.

Ramakers, P.K.J. & van Lieburg, M.J. (1982) Start of commercial production and introduction of *Amblyseius mackenziei* for control of *Thrips tabaci* in glasshouses. *Meded. Rijksfac. Landb. Gent* **47**: 541–545.

Smith, C., ed. (1966) *Insect Colonization and Mass Production* Academic Press, London & New York. 618 pp.

9 PRODUCTION AND COMMERCIALIZATION OF PATHOGENS
S.G. Lisansky

Research into the practical use of fungi which kill insects is being carried out by many scientists and horticulturists throughout the world. Other alternatives to chemical pesticides are also under investigation, including novel pest control systems based on parasitic and predatory insects, fungi, bacteria, viruses, protozoans, nematodes and pheromones which modify the behaviour of insects for Man's benefit. However, most of these novel systems have not yet been exploited in horticultural practice on a commercial scale. This is due to the series of sometimes lengthy and sometimes expensive steps which must be carried out before new pest control systems can be offered to the grower (Figure 42).

9a **STEPS LEADING TO COMMERCIALIZATION** Once any new system has been identified and characterized in the laboratory, the following steps must be completed before it can be successfully commercialized:

1. PROCESS DEVELOPMENT A process must be developed which can be carried out on a large enough scale to ensure that an adequate amount of the material can be made. This process must be sufficiently reliable to provide a product which is both safe and effective. In addition, the production costs must allow manufacturers to make a profit.

2. PRODUCT DEVELOPMENT Many systems for controlling pests are successful in the laboratory. However they fail when tried in the greenhouse or in the field. This is one of the many problems which must be solved by 'product development'. Products must be manufactured and formulated in a way which makes them stable for the longest possible time, as convenient as possible to use, and as immune as possible to 'user abuse', i.e. the failure of many people to store or use the product as directed and then to blame the product for poor performance.

Recommendations must be developed for the use of the product in actual horticultural or agricultural practice. Normally these recommendations cannot be so novel or so complex or require such unconventional equipment that people refuse to use the product. Often the recommendations are developed in collaboration with the appropriate

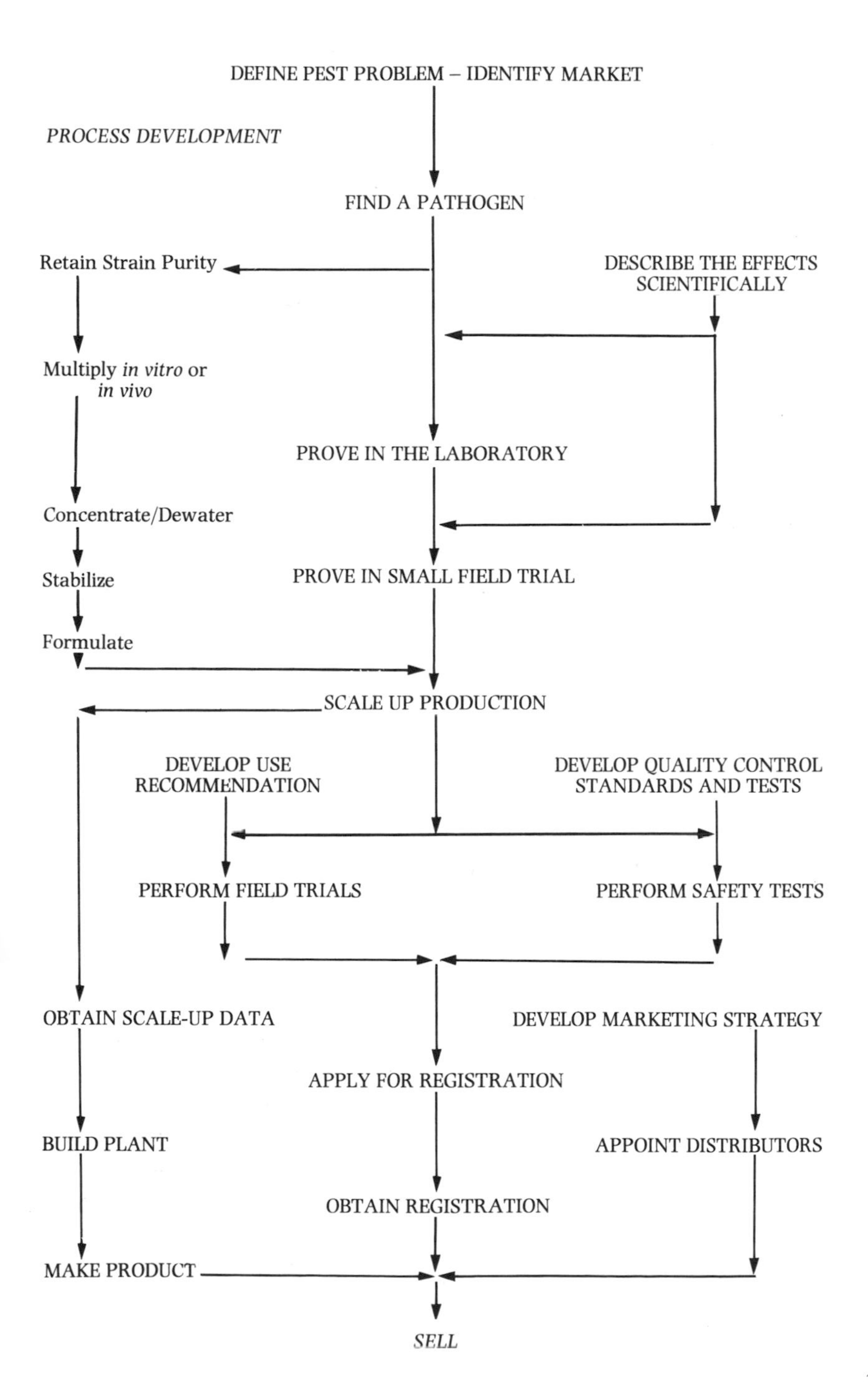

FIGURE 42 Commercialization of pathogens.

advisory service to ensure that they will be accepted by the trade.

For novel products, testing is necessary to establish that they are safe to use; in addition, quality control tests must be devised which will ensure that every batch will be safe and effective. Once any testing is completed and quality control protocols developed, the appropriate government authority must be approached for permission to sell the product; at present, this must be done individually for every country in the world.

Once a product has been sold, it is often helpful to visit growers who have used it, either to confirm its success or, if it has failed, to investigate and determine the cause. Further development work can then be carried out to find ways of avoiding product failure in the future.

3. MARKETING AND SALES Once made, the product must be sold and used if all the effort spent in development is to be worthwhile. There are many obstacles to selling alternative pesticides, especially fungi. Users are unfamiliar with how to use these types of product and how they work. Marketing requires extra efforts both to familiarize growers with the new product and to provide a substantial back-up service for the product.

The preceding section outlines problems which must be solved for any novel pest control system. We shall now discuss the particular problems which must be solved during the development and commercialization of fungal pathogens for the control of insects.

9b **PROCESS DEVELOPMENT AND PRODUCTION** Before one can sell a live micro-organism as a pest control agent, a reliable method of production must be developed which yields large quantities for which a product specification can be drawn up.

1. **Organism storage** The first problem in producing a micro-organism is storing it in a way that ensures the retention of desirable features. The most obvious feature to be retained is pathogenicity; the second is productivity in terms of yield in the commercial production process. (This is typically measured either by total biomass, or by number of infective propagules, i.e. spores or mycelial fragments, produced per litre-hour of fermentation time.) Many micro-organisms are known to lose desirable features either on storage or after repeated sub-culturing. Amongst insect pathogens, *Beauveria bassiana* has been reported to have reduced virulence after sub-culturing whereas both *Verticillium lecanii* and *Metarhizium anisopliae* are reported as being undiminished in virulence after many passages. Organisms which lose virulence may sometimes be restored to their former potency by passing them through their normal host, i.e. the target insect; however, such a technique would be cumbersome as a routine part of a production process and

always presents the risk of contamination. Therefore the problem is often avoided by storing a large number of aliquots of a single spore isolate in a deep frozen or freeze-dried condition. Samples are checked periodically to make certain that virulence and productivity have not diminished with time.

2. **Fermentation method** The standard method of production of micro-organisms is the process of fermentation. There are many types of fermentation; the two commonest are 'submerged' and 'semi-solid'.

Submerged or 'deep-tank' fermentation is, as the name implies, the growth of micro-organisms in a fully liquid system. There are a number of advantages to fully liquid systems which include the ability to hold temperature and pH constant, the ability to pump large quantities of air into the system and disperse it by means of stirring impellers, and the ability to generate reasonably homogeneous conditions to maximize the growth of the micro-organism.

Despite the many advantages of submerged fermentation, some fungi will not yield a satisfactory product by this technique. Semi-solid fermentation offers an alternative in which the fungi grow primarily on the wet surface of a solid material, often some form of processed cereal grain to which nutritional adjuvants have been added, though attempts are made frequently to use 'waste' materials or substrates of low value, such as straw. This allows fungi to grow in conditions more similar to those found in nature; spores, the infective propagules by which the fungus survives and infects insects, are produced in the air and are consequently more durable. Semi-solid fermentations are relatively easy to develop on a small scale (see Figure 43 of a shake flask). Scaling them

FIGURE 43 Mass production of pathogens – shake-flask culture.

FIGURE 44 Mass production of pathogens – large rotary fermentor.

up to the sizes necessary for commercial product (Figure 44) presents numerous problems; aeration becomes a major difficulty as the volume of a semi-solid mass increases more rapidly than the available surface area. This requires either a very large area of relatively shallow substrate, e.g. on trays, or in a vessel which can agitate or tumble the substrate. On any scale, trays are very difficult to sterilize and keep sterile. The development of large vessels for semi-solid substrate fermentation requires the invention of a number of techniques or pieces of equipment for:

(a) keeping the substrate friable after sterilization (its tendency is to set solid when it cools, rather like oatmeal)
(b) inoculation with the desired fungus without contamination
(c) aeration and agitation during the fermentation
(d) drying the material prior to opening the fermenter in order to avoid contamination, etc.

3. **Medium development** Whichever type of fermentation is chosen, nutrients must be provided so that the micro-organism can grow. Which nutrients are chosen will markedly affect how fast the organism grows, how much is produced and, often, how infective the final product is. Nutrients to be provided include a carbon source, e.g. glucose or molasses, a nitrogen source, e.g. soy bean meal or yeast extract, and an assortment of other molecules needed in smaller amounts. Media can be 'defined', in which case the precise nature and quantity of every nutrient is known, or, alternatively, they can contain ingredients of an indeter-

minate and occasionally variable nature, e.g. molasses; most commercial media are of the latter type.

4. **Downstream processing** (a) FROM A SUBMERGED FERMENTATION When the fermentation is complete, the product can be pumped, harvested and formulated. Harvesting of the 'active ingredient', i.e. the live microorganism, can present a problem due to its relatively low concentration in the broth and, usually, its sensitivity to the conditions, such as high temperature, created during processing. If the product is to be sold as a liquid, it can be harvested by centrifugation or some form of filtration which increases the concentration of solids. Adjuvants can be added both to assist in this concentration step and to prevent degradation or spoilage of the product. If the product is to be a dry powder or granule some means must be found of drying it without harming the microorganism so that the final product achieves the desired potency. Such means of drying include spray-drying, freeze-drying, vacuum drying and fluidized bed drying; in addition, a variety of types of centrifugation or filtering may be employed, with or without flocculation or filtering aids prior to any drying step.

(b) FOLLOWING SEMI-SOLID FERMENTATION After fermentation and drying, the product must be processed into a powder by comminution of some kind. Achieving this without an unacceptable loss of activity is still an unsolved problem.

In the development of production methods, economic considerations become very important. Media cannot contain ingredients which are too expensive, or which vary too much in availability or quality; each batch which fails to meet specifications is a waste of time and money. Fermentation is costed on a 'litre-hour' basis. The longer something takes, or the lower the yield, the more expensive it becomes. For some potentially useful insecticidal fungi, the main inhibition to their further development is the lack of an economic production process. This is true for *Erynia neoaphidis* and could well be true for *Aschersonia.*

9c **PRODUCT DEVELOPMENT** 1. **Formulation** Although formulation is performed at the end of the production process, formulation development is one of the problems solved during product development. It has two primary purposes:

(a) to stabilize the product so that it can be stored until needed.

(b) to make it easy, convenient, safe and reliable to use.

Unformulated fungi frequently do not live very long. *Verticillium* from a submerged culture may survive only a few hours; formulated products based on *Verticillium* are reported as losing no activity for as long as 1 year.

In addition, formulation makes the product easy to wet and dilute for spraying, gives it the ability to stick to leaf surfaces or host insects, allows it to survive for several days on its own before infecting an insect and provides nutrients and moisture-holding ingredients to promote rapid infection.

2. **Use** Although the potential of fungi as insecticides may be fully demonstrated in the laboratory, their use in commercial horticultural situations requires that directions be given to maximize the chance that the fungus will be successful. Use recommendations are, of course, also needed for chemicals and many excellent chemicals will not work if misapplied. However, fungi are, at present, particularly sensitive to mishandling so extensive experimentation must be done to establish both the optimum conditions for use and the extremes beyond which successful control is very unlikely.

3. **Specification** At a certain stage of its development, the product should cease to vary. People involved in making, testing, approving, or using the product must be working on the same thing. The specification defines the nature and minimal purity of the product. In order that the product will not vary over time, the production process must be well tested and the formulation reasonably settled. Changing the product specification may result in having to retest the product for both safety and efficacy, which can be a lengthy and expensive process.

4. **Testing** Once specified, the product must be tested for safety. Tests required on animals can include 'acute toxicity' to rats using various routes of administration, 'dermal and eye irrigation' studies, and tests of the product's allergenic potential.

Agrochemicals are now tested very extensively; fungi and other micro-organisms are subject to fewer, less expensive tests than those required for chemicals, due to their inherent safety. Nevertheless, the estimated cost of performing the minimum tests required by the United States Environmental Protection Agency is $80 000–120 000.

These tests must show that the product is safe before it is allowed to be sold; if they do not prove the product is safe, the money spent on testing and all the previous development work is lost.

5. **Registration** If the product is shown to be safe, it must still be registered in all the countries where it might be sold. Each country in the world acts individually on these registrations. Some may require different, or additional, safety tests. Each will require proof of the efficacy of the product; they will require the trials to be performed in that country and to be under the supervision of local personnel and also that the cost of

these trials is charged to the manufacturers or distributors, whether the trials succeed or not.

Many countries have a stated desire to replace some of the agrochemicals used with 'environmentally softer' products such as microbial pesticides. Nevertheless, when presented with a registration application on which action is required, the bureaucrats in these countries may prevaricate and delay, which can 'kill' a product just as surely as poor performance. A reasonable development plan allows at least 2 years to complete a registration, even after most tests have been performed.

6. **Quality control** If the product is shown to be safe it is necessary to establish methods of ensuring that each batch is as safe and effective as the previous one. These Quality Control Standards should be set to measure:
(a) efficacy, probably by bioassay against the target insect.
(b) number of infective propagules, as a standard for the label.
(c) microbiological purity and absence of dangerous contaminants.
(d) mammalian safety, a small extra test on mice as insurance.
(e) physical characteristics of the product.

At this stage, with confidence in the product and its method of production, the building and equipping of the plant which would manufacture the final product has to be planned; patents and trademarks to protect both inventions and product must also be sought.

9d **MARKETING AND SALES** The ultimate test of any product is whether it can be sold at a profit. During all the development outlined above, the marketing department would have been planning how to sell the product and to whom.

However, before the development even starts, a market feasibility study must be carried out:
(a) to determine whether there is any need for such a product,
(b) to assess the amount that may be sold,
(c) to assess the competing products, whether chemical or biological,
(d) to conduct an analysis on which a decision can be made.

Having decided to proceed, the marketing department:
(a) develops the strategy to be followed at the time of commercial introduction and thereafter.
(b) analyses distribution networks and appoints agents who are best suited to handle the new product.

For biological products, it is important to appoint a firm who know their market sector well and provide it with a high level of back-up service; this means that agents must be either fairly big or fairly local. It is desirable to appoint agents who have experience with registration and good contacts within the advisory services.

These actions are pre-conditions for sales. Once in a position to start sales, advertising and publicity become important as means of informing potential customers that the product is ready.

Sales Selling the product is the most difficult task of all. Very few products are so clearly superior to every competing product that they sell themselves; and when they are superior and become the market leader, they become the target which others try to imitate or better. Microbial pesticides are rarely cheaper than competing chemicals; they frequently will not give the quick and complete kill that chemicals give.

Customers may use microbial pesticides because:

(a) the pests are resistant to all available chemicals.

(b) they fit into an integrated programme using other beneficial organisms.

(c) the grower may believe them less toxic to his plants.

(d) the grower may believe them safer to himself or the environment.

(e) the government bans alternative control measures (e.g. DDT).

(f) the crop is in a 'market niche', i.e. in a sector where the chemical industry cannot justify developing a product because the market is too small or too low in value.

After sales service Every product has some dissatisfied customers; microbial insecticides may have more than average because they are novel and unfamiliar and their mode of action is slower than most chemicals. An important aspect of marketing is to visit as many customers as possible, finding out where and why the product has worked poorly and assisting the grower in subsequently achieving good results. Visiting customers will also help to document where the product has proved effective and may yield ideas for future improvements. Ideally, product development should be a process of continual improvement.

10 THE ADVISORY PROBLEM
H.J. Gould

10a **HISTORICAL BACKGROUND** Since the early 1950s, pesticide resistance in the major pests of protected crops has been an increasingly important problem and a serious limitation to the successful control of pests on such crops as cucumbers, tomatoes and chrysanthemums.

In the UK, the problem first occurred with the red spider mite (*Tetranychus urticae*) which became resistant to azobenzene in 1950, followed by resistance to parathion in 1952 and to the majority of organophosphorous pesticides by 1960. Pesticides such as tetradifon and dicofol had a relatively short commercial life as fully effective materials and resistant strains are now widespread, although there is, as yet, no evidence of resistance to dienochlor or cyhexatin. Other important pests, such as glasshouse whitefly (*Trialeurodes vaporariorum*) and the peach-potato aphid (*Myzus persicae*), are also now resistant to organochlorine, organophosphorus and some carbamate pesticides. In 1983, resistance to the recently introduced synthetic pyrethroids was confirmed in two strains of whitefly from protected crops in the UK.

When the resistance problem was first recognized, the control policy usually adopted was to change to another pesticide with a different mode of action or, where available, to use mixtures of pesticides (e.g. tetradifon/dicofol). This strategy was initially successful but depended on a regular supply of new pesticides. In many situations, the choice of pesticide was limited by problems of phytotoxicity on high value crops, the lack of 'cleared' pesticides for relatively minor crops which were not commercially attractive to the pesticide industry and, more recently, by the smaller number of new compounds being developed.

Another constraint to the successful control of pests on protected crops during this period was the pressure, for reasons of costs and convenience, to move away from HV spraying to other, often less reliable, labour-saving techniques such as smokes, ULV spraying or fogging, all of which generally required more frequent applications to maintain adequate pest control.

10b **INTRODUCTION OF INTEGRATED CONTROL TECHNIQUES** By the mid 1960s, control of the major pests on cucumbers and chrysanthemums in particular could be achieved only by the use of costly intensive programmes of pesticides and, on cucumbers, it was not unusual for fifteen or more applications of various acaricides to be used during the growing season in an attempt to keep red spider mite in check.

Conditions were therefore ideal for the introduction of new techniques based on integrated pest management programmes, using parasites, predators and pesticides. The programmes for the control of the pest and disease complex on cucumbers, tomatoes and chrysanthemums which have been described in earlier Sections were largely developed to solve the resistance problem.

Supplies of *Phytoseiulus persimilis* and *Encarsia formosa* have been available to growers in the UK from a number of commercial suppliers since the early 1970s. In the early development of the technique, supplies of parasites and predators were provided by a number of growers producing the material as a 'side-line' to their horticultural enterprises. The material was, therefore, often of uncertain availability and quality. This caused some difficulties for growers pioneering these techniques and control was consequently often unpredictable. As the demand for natural enemies increased, and it became clear that there would be a definite market, a number of larger specialist producers became established and a regular supply of good quality material is now available from suppliers in several countries including the UK, Netherlands, Finland, France, Norway and Sweden.

10c **GROWER/ADVISOR PHILOSOPHY** The commercial development of the new techniques raised a number of questions for the grower and advisor. The grower, previously expecting 'instant' results from the use of pesticides, had to understand that, with biological agents, control takes place over a much longer period (e.g. 6 weeks for red spider mite control and some 16 weeks for whitefly), the interaction between 'predator and prey' often continuing for the life of the crop. Also it was important for the advisor to convince the grower that he could safely accept small numbers of pests on the crop, below the economic threshold level, which were actually essential to maintain the biological interaction.

The early programmes were frequently based on a very novel proposal in which the pest was actually introduced on to the crop at a predetermined level in order to set up a predictable interaction with the beneficial agent introduced slightly later. Although this method has undoubtedly been shown to give the most reliable results for the control of red spider mite and whitefly on cucumbers and tomatoes, it was not popular with growers and modified programmes depending on natural infestations were developed and are now almost invarably used.

At an early stage, it became obvious that growers would themselves have to become closely 'integrated' with the programmes being used in their nurseries and that careful monitoring of the establishment of parasites and predators was necessary at regular intervals. To do this successfully, they would require a basic understanding of the biology of the pests and natural enemies and a knowledge of various critical stages of the parasites and predators being used. In the UK, both the Agricultural Development and Advisory Service (ADAS) and the Glasshouse Crops Research Institute (GCRI) played an important role in producing technical literature and in giving courses to growers and their staff. With the commercial development of the techniques, useful instruction leaflets are now provided by the suppliers.

Although the results obtained using these programmes were generally satisfactory, failures sometimes occurred. A number of factors were found to be responsible for these poor results. The main problem was the delay in the introduction of the parasite or predator which allowed pest levels to exceed those at which biological control could be achieved in a commercially acceptable period before excessive crop damage took place. Excessive trimming or deleafing of cucumbers and tomatoes during the establishment phase of the whitefly parasite (*Encarsia formosa*) frequently caused control failures as large numbers of parasitized scales were removed with these lower leaves. Lack of hygiene in the control of weeds around the greenhouses early in the season often provided a reservoir for large numbers of whitefly which could infest the new crop at an early stage before the parasite was well established.

Another major problem in the early development of the technique was the unexpected 'flare up' of minor pests such as thrips, capsids, leafhoppers and leaf-miners which were able to colonize the crops in the absence of the routine insecticide programme. Growers using unsuitable pesticides for the control of these pests in the presence of a biological control programme for the major pests would inevitably upset the natural enemies and lose control of the situation. Control strategies for these minor pests have been fully discussed in Section 3.

10d **SCALE OF USE OF INTEGRATED/BIOLOGICAL CONTROL IN EUROPE** Since 1975, annual surveys have been made by ADAS of the extent of use of biological control programmes on both tomatoes and cucumbers (Gould, 1980). The information collected included details of the area of tomatoes and cucumbers over which *Phytoseiulus persimilis* and/or *Encarsia formosa* was applied. The results of the surveys from 1975–80 are shown in Table 10. Additional data from the ADAS pesticide usage survey (Sly, 1981) is also presented for 1981.

The surveys have shown that there was a very rapid uptake in the use of biological control on cucumbers from 1976, particularly for the control of red spider mite, but a more gradual introduction of the use of the parasite to control whitefly on tomatoes. In recent years, progress has been slower, with little increase in the use of *Encarsia*; this coincides with the introduction of the synthetic pyrethroids which, at least temporarily, afford effective control of whitefly on protected crops.

TABLE 10 Use of biological control on cucumbers and tomatoes in the UK as percentage cropped area

	Cucumbers		Tomatoes	
	P. persimilis	*E. formosa*	*P. persimilis*	*E. formosa*
1975	7	10	2	3
1976	45	27	8	16
1977	47	43	15	27
1978	72	36	20	31
1979	63	40	26	40
1980	74	46	11	39
*1981	67	46	17	30

*Based on Pesticide Usage Survey, Sly (1983).

In addition to the use of biological control on cucumbers and tomatoes, the technique is also used on some of the larger year-round chrysanthemum nurseries and by a small number of ornamental and foliar pot-plant producers. There is also a very limited use of *P. persimilis* for the control of red spider mite on protected strawberries.

Although data have been presented only for *P. persimilis* and *E. formosa*, where appropriate, the integrated control programmes make full use of all other available agents, including *Verticillium lecanii*, *Bacillus thuringiensis*, leaf-miner parasites (*Diglyphus sp.* and *Dacnusa* sp.), Thripstick® and various selective pesticides (e.g. pirimicarb).

Information on the extent of use of biological control in the rest of Europe was presented by Woets and van Lenteren (1983) and a modified summary of their data is presented in Table 11.

TABLE 11 Use of biological control on protected crops in Europe 1982 as percentage of cropped area

Crop	Country	Crop area ha	% Crop area *P. persimilis*	% Crop area *E. formosa*
Cucumber	*Denmark	55	75	—
	Finland	60	83	5
	Netherlands	725	65	3
	Norway	25	80	4
	Sweden	58	52	—
Tomatoes	Bulgaria	700	—	1
	*Denmark	115	—	50
	Finland	180	1	11
	Netherlands	2100	1	23
	Norway	50	20	10
	Sweden	64	—	39
Sweet peppers	Netherlands	250	16	—

*1978 data, Berendt (1980).

The use of *P. persimilis* on cucumber and *E. formosa* on tomatoes follows a similar pattern to that in the UK but there has been little development of the use of *E. formosa* for the control of whitefly on cucumbers because of unpredictable results.

The advisor in integrated control is basically concerned with 'selling' confidence. To undertake this effectively he must establish an effective relationship with a grower who enjoys a reputation for production efficiency among other growers of that crop. On the chosen holding, confidence is built up by advancing steadily from small experiments to adoption of the methods over the whole nursery. Through 'open' days and discussions with the host grower, others can often be persuaded to adopt the methods. Advisors must recognize that scale is the deciding factor and this puts officials in a difficult position as their administrative seniors fear claims for failure.

In the case of pesticides, official bodies can pass the responsibility and risk to chemical companies but with biological control, where a 'method' rather than a 'product' is being 'sold', they are more vulnerable. Hence, it is vital to use outstanding growers to 'publicize' and 'sell' the technique to less adventurous colleagues.

REFERENCES

Berendt, O. (1980) Trends in biological control in glasshouses in Sweden and Denmark. *Bull. IOBC/WPRS Working Party on Integrated Control in Glasshouses, Vantaa* **3** (3): 11–16.

Gould, H.J. (1980) The development of biological control of whitefly and red spider mite on tomatoes and cucumbers in England and Wales. *Bull. IOBC/WPRS Working Party on Integrated Control in Glasshouses, Vantaa* **3** (3): 53–58.

Sly, J.M.A. (1983) Pesticide Usage on Glasshouse Crops in England and Wales 1981. *MAFF Survey Report* No. 14.

Woets, J. & van Lenteren, J.C. (1982) Areas with application of beneficial insects in greenhouses in Europe. *Sting* **6** August.

11 THE ECONOMIC EQUATION
N.W. Hussey

11a **THE STATE** As part of a continuous programme of research support to their horticultural and agricultural industries, all Western European countries have financed the developments outlined in other chapters. In general, however, these developments have occurred through the enthusiasm of individual research workers rather than by a considered strategy. Indeed, there has frequently been a conflict of ideals between research teams and their administrative masters, as the latter usually claim to see a clear distinction between research and development. In practice, the field of integrated control can be successfully exploited only when there is a continuum of interest and involvement between research and extension workers and the growers on the nurseries where ideas are tested.

Whatever these conflicts of interest, the politicians have consistently joined the bandwagons of safety and environmental protection. All too often, these laudable goals have led to a negative pressure on the pesticide companies and their developmental programmes. It is essential that society recognizes that pesticides have, and will always have, a dominant role in crop protection. The biological approach offers an opportunity to avoid resistance and so extend the effective commercial life of products which have cost as much as £15 million to bring on to the market.

The greenhouse environment makes only a minute contribution to the national environment, though the success of the biological approach there has provided an important stimulus to its use outdoors. Indeed, the current advances in the use of naturally occurring predators in intensively grown cereals, which arose as a direct stimulus from greenhouse work, may well prevent the trend towards 'insurance' control of aphids on vast acreages of wheat and barley.

Whatever the value of chemicals, there is no doubt that their use within greenhouses creates serious hazards to operators who apply them. Their substitution by completely safe natural methods is, therefore, welcomed by both governments and growers.

These advantages have, of course, been gained only after considerable investment. In the UK, we calculate that the first successful integrated programme on cucumbers was developed at a cost of £500 000. As will become clear, the economic advantage to be gained from the use of these programmes over several years will return this investment with strikingly positive cost/benefit ratios.

When methods of mass-rearing of beneficial insects were under development at the GCRI, the staff involved believed that a rearing unit would be established at the Institute. However, no acceptable political solution could be negotiated between the different interests involved and so the field was left open to private enterprise. This unexpected situation led to unnecessary research and development expense as all the preliminary studies were directed towards a single, large 'factory-type' unit. These methods were quite unsuitable for the greenhouse production systems ultimately adopted by several growers and which now constitute the main form of production.

11b **THE PRODUCER OF NATURAL ENEMIES** In the Netherlands, only one individual set up in business and the very large concentration of local customers enabled Koppert BV to become the largest and most heavily capitalized facility. In the UK, on the other hand, several units were established to serve a smaller and more widely dispersed market. At first, this presented no problem as the several owners continued to conduct their greenhouse business but, as the competition and the requirement for high quality products increased, the lack of cash flow and hence investment capability enabled Koppert to assert a commanding position.

When the number of UK producers effectively declined to two, they sought to widen their product range by selling new natural enemies which had received little attention from official research workers. This development emphasized a basic weakness in the biological control industry – namely a desire to provide a complete alternative to the chemical package. There is little doubt that only a few growers are prepared to accept this philosophy, so that it has proved difficult to make the diversification really profitable.

These problems have been further emphasized by the marketing of microbial products where the manufacturer has nominated a sole distributorship. These products require a longer development period prior to marketing than is traditional with chemical pesticides and hence the inevitably increased costs need to be sheltered by a good cash flow from other products. To date, none of the companies is in this favoured position, though, interestingly, the dominance of Koppert is being challenged by Bunting & Sons who also operate a successful tomato nursery and sell a number of other requisites to growers.

There is no doubt that larger markets are required to enable these rearing firms to break out of the financial straightjacket which restricts diversification.

One such extension has already occurred, namely into outdoor strawberries, and had the interesting effect of reducing the market price of *Phytoseiulus* by 50%, so that the large numbers of predators required could be purchased at a realistic cost for that crop. In the future, a real possibility lies in the massive areas of temporary plastic coverings used for vegetable production across the Mediterranean from Morocco to Macedonia. The area involved, more than 20 000 ha, could offer a cash flow of £2 million even if a market penetration of as little as 10% were achieved. However, as pointed out elsewhere, the existing methods of integrated control cannot be easily translated to the very different cultural situation in the Mediterranean. The rewards are great and progress in this area could be more profitable than attempting to seek greater returns by mass-rearing other beneficial insects, such as *Trichogramma* for European corn-borer.

Since successful integrated control is largely a matter of confidence, several UK companies attempted to provide expertise to assist growers to operate their biological control effectively. Although a small island, the dispersal of the greenhouse industry is sufficient to make the conduct of such a service uneconomic unless operated in connection with some other service to growers. One company linked the biological control service with one for methyl bromide fumigation, which demanded that their operators were in any given area for 24 hours. Another attempted to service the programmes with qualified entomologists. Koppert has usually operated his business through firms already selling pesticides to growers.

Whatever formula has been adopted, the results have not always been entirely satisfactory, as the staff involved have either not been especially well qualified to deal with this complicated and rapidly changing field, or have been unable to reach nurseries sufficiently frequently. There is no doubt that the science of integrated control will not attain respectability and efficiency until the major pesticide companies enter the field with determination and use their considerable resources of technology and qualified staff to market 'control' rather than cans of chemicals. Such a move cannot be regarded as imminent but the need to protect profitable pesticides from resistance problems seems certain to move industrial opinion much closer to this decision by the end of the next decade.

11c **THE GROWER** Perhaps the first important, but unquantifiable, advantage to come from biological control was the extra time which growers found at their disposal following its use. During the summer months, pesticides must be applied at the end of the normal working day to avoid the phytotoxicity which commonly occurs where the sun's rays play upon chemical droplets on leaf surfaces. These applications were both costly and inconvenient as the operators involved, while paid at overtime rates, were kept away from the family circle by the unsocial hours. Indeed, one small grower who applied his own pesticides regarded biological control in a specially favourable light as it enabled him to restart his social life by attending the theatre etc.!!

There is much interest in the comparative economics of chemical and biological control but there are no sweeping generalities that can be made, as each crop presents its own problems which must be considered separately. However, there is one interesting and consistent difference between the systems, namely that biological methods usually have minimal application costs though the cost of the products is high whereas, with pesticides, the opposite situation occurs.

When the first integrated programmes were developed, they were designed to cost only half the sums incurred in the full chemical programmes which often involved more than 20 sprays per crop. This target was met and was undoubtedly a major factor in the initially rapid uptake of the technology. However, as red spider mite, the major pest involved, came under control, less pesticides were required and the direct cost comparison was less advantageous to natural control. Even now, 15 years later, integrated programmes cost less than the purely pesticide alternative. At a major chrysanthemum nursery in 1983, the comparative costs per 1000 plants were £1.12 and £1.34 respectively.

It is important to remember this changed situation when considering this economic parity for one is really buying an insurance against the selection of resistant strains which, if it occurred, would make costs of the purely chemical programme rise rapidly.

Another aspect of the economics of biological control concerns the apparently high costs of individual biological products. Certainly, if viewed in isolation, they are considerably more expensive than chemical products with which they compete. However, within an integrated programme of which they are a vital component their costs are hidden within that for the total programme and so become economically acceptable. A good example is provided by the microbial Vertalec® which costs about 4 × as much as pirimicarb but, by supplanting all other aphicidal applications and allowing the predator to perform effectively, the integrated package becomes less costly than the full pesticide programme.

On cucumbers, the largest UK cucumber-producer was paying about £20 000 (£300/ha) for pesticides without considering application costs, a bill which was reduced to almost nil once *Phytoseiulus* and *Encarsia* came into use. The cost of rearing these natural enemies in an 'in-house' production unit, including technical staff, was about £400/ha in 1978.

Perhaps the most important reward from the use of biological control is the increased crop which follows the cessation of pesticide applications. This phytotoxicity is not accompanied by symptoms of damage but the solvents and other adjuvants apparently affect the fruit set of glasshouse vegetables. Addington (1966) demonstrated a crop increase of 25% on his 7 ha while large growers of both cucumbers and tomatoes who have efficient crop-recording schemes have also claimed between 10 and 15% increases. Bearing in mind that these crops are valued at about £125 000/ha the reduced pesticide costs are coupled with a bonus of up to £12 500/ha!! Of course, in the current changed circumstances of lighter pesticide programmes, these enormous economic advantages are less obvious. Nevertheless they have made a significant contribution to meeting the vast increases in fuel costs over the past decade.

Even flower-growers have reported significant increases in the proportion of top quality produce so that all growers must accept that biological control affords them a very real commercial advantage in return for reduced crop protection costs.

The widespread public interest in 'natural food' has provided certain growers with an opportunity to increase the value of their produce by selling pesticide-free fruits at a premium. This form of marketing is usually operated by direct marketing to specialist outlets of natural products. The current health phobia amongst the general public of Western Europe would seem to ensure a steady increase in this market, however irrational the fear that conventionally marketed products bear undesirable chemical residues. However, since many pesticides are used within a few days of harvesting, there is a danger that some unscrupulous operators will succumb to the temptation of using any authorized product to guarantee a clean harvest.

As with many technologies there are other, less quantifiable, advantages. It is all too easy for nursery staff to treat their tasks as merely routine with a consequent lack of sharpness in the conduct of their duties. The management has to continually counter these attitudes of mind since they may lead to the lack of recognition of cultural problems which can only be economically dealt with if acted upon at the first sign of trouble. Nursery-workers rapidly take an interest in biological control if adequately briefed and hence become accustomed to paying attention to details which they would not, hitherto, have noticed. Those nurseries where this change in attitude has developed are those on which considerable trouble has been taken in their training and awareness. It is salutory to remember that the enormous use of biological control in China depends on the informed co-operation of the peasants. Managers would welcome such an effective work force which can be a direct result of a complete implementation of the integrated control strategy.

REFERENCES

Addington, J. (1965) Satisfactory control of red spider mites on cucumbers. *Grower* **66**: 726–727.

12 INTERNATIONAL CO-OPERATION N.W. Hussey

The stimulus to biological control which has developed since the 1939–45 war was reflected by the formation of a Working Group of OILB which first met at Naaldwijk (the Dutch Glasshouse Vegetable Research Station) in 1970. Nine delegates from 8 countries attended the meeting and 11 papers were presented. At that time *Phytoseiulus* was used on a small scale in the UK, the Netherlands and Austria, whilst *Encarsia* introductions had begun in Canada and the UK. At that meeting, particular attention was directed towards the harmonization ot chemicals with biotic control methods. It was suggested that integration could be achieved by: (a) applying pesticides only when natural enemy populations would not be seriously harmed, either because they were protected by the foliage or had almost achieved control, (b) using 'spot' treatments and (c) using truly selective pesticides.

Three years later, Dr Bravenboer, convener of the Group, arranged another meeting at Littlehampton (UK). This was attended by 22 delegates from 10 countries and considered 17 technical papers. By this time there were 5 companies producing natural enemies (UK 2; Netherlands 2; Finland 1). While some attention was given at the meeting to production and introduction schemes, much of the discussion centred on aphid control with parasites, predators and pathogens.

By the third meeting of the Group at Antibes in 1976, the attendance had increased to 25 and reports of biological control applications were presented from 11 countries. Particularly important reports were presented by Finland on *Aphidoletes*, by the Netherlands on the selection of an organophosphorus-resistant strain of *Phytoseiulus* while the threat of *Thrips* was recognized for the first time.

This series of OILB meetings have always been regarded as 'report and discussion' sessions designed to stimulate workers in different countries to try out ideas and techniques developed elsewhere. One or two attempts were made to draw up experiments that could be conducted simultaneously in different countries but, with the exception of work on the 'pest-in-first' methods, members found it difficult to inject such international projects into their national programmes. However, in 1978, a special meeting was held at Littlehampton to discuss the strategy to be adopted by the groups at Naadlwijk and GCRI undertaking research designed to control *Thrips tabaci*. This gathering agreed that the Dutch group should pursue their interests in predatory control while at GCRI attention should concentrate on selective chemical control. Such division of interest ensured that no overlap of effort occurred and, in the event, led to the development of both *Amblyseius* and Thripstick® to the stage of commercial use. We are now waiting for the market place to pursue these developments and decide which should be preferred, because it is unlikely that both can survive commercially.

In 1979, the Working Group met at Vantaa (Finland) where much interesting work had been conducted on aphid predators and where the firm, Kemira Oy, had launched a number of commercial techniques of biological control.

The meeting attracted 36 participants from 12 countries and reported on commercial developments in 11 countries. Interest was expressed in biological control of tomato leaf-miner based on research at the University of Leiden in the Netherlands. It should be remembered, however, that leaf-miners had already been successfully controlled by parasites on year-round chrysanthemums in small commercial trials in the UK. At this meeting, it was apparent that, although there was more basic research than ever before, the area of protected crops on which the techniques were commercially applied had not significantly increased.

There is no doubt that this is due to an over-concentration on biological control rather than the vitally important, but less attractive, field of practical integration. In an attempt to stimulate more work in the latter challenging field a small specialist meeting was held in Naaldwijk in 1981.

Only two subjects were discussed, namely *Thrips* control and investigations designed to improve control of whiteflies by *Encarsia*, which had been hampered by the trend to low-temperature growing as a contribution to reducing oil costs. This development had thrown into relief the importance of a fuller understanding of the environmental factors which influence the flight of adult parasites. As a result of these discussions an important series of experiments were conducted in the Channel Islands during the following winter and reported to the fourth full meeting of the Group in Darmstadt in 1982. 42 workers from 14 countries attended this meeting which was especially significant as several colleagues from Japan and the People's Republic of China attended for the first time.

Among the important matters discussed were the relative merits of the entomophilic fungi, *Verticillium lecanii* and *Aschersonia aleyrodes*, for whitefly control as workers in the Netherlands and the UK had made several contributions to this interesting subject. An especially valuable discussion centred upon the potential role of adhesive yellow traps in integrated pest management programmes for tomatoes in Sicily and on chrysanthemums in the USA. In practice, it would appear that environmental conditions play a major role in the efficiency of traps, by increasing or otherwise, the flight activity of adults.

The work on whitefly control after planting of the crop in autumn, followed by night temperatures of 12°C throughout the winter, showed that it was essential to ensure that no whiteflies were present on the plants when the new night temperature regime (12°C) commenced. A careful study on the preplanted seedlings revealed that reduced rates of oxamyl and aldicarb could achieve this objective without phytotoxicity.

International research effort and collaboration over the past 15 years has enabled fully integrated programmes to be mounted but the immediate future effort must concentrate on demonstrating the important advantages to the grower so that more can be attracted to take up the methods and the protection of the treated areas be increased to a high, uniform percentage. Paramount is, of course, publicity and commercial demonstration.

Publicity is largely a matter for national crop protection services but it is important for grower organizations to bring to a wide audience the considerable economic potential of the biological approach, especially in the highest value crops. Practical experience also shows that unheated plastic structures provide many opportunities for the techniques on new crops such as strawberries and carrots.

The other aspect of publicity is, of course, demonstration and we hope to stimulate participants at further meetings to mount fully integrated programmes in their own countries. These programmes are available from the GCRI and Naaldwijk, for ornamentals from the Dutch station at Aalsmeer and the important French centre at Antibes, and from the Agricultural Development and Advisory Service in the UK.

The next meeting of the Group is scheduled for the summer of 1984 in Sicily, where it is hoped to set up development trials in some of the temporary plastic structures widely used for growing vegetables in the Mediterranean region.

Before concluding this survey of international co-operation, tribute must be paid to Dr Bravenboer of Naaldwijk, who founded and stimulated the Group for more than twelve years. Now that knowledge has reached the current level the future challenge is to encourage members to help set up development/demonstrations in order to influence national grower associations. This goal may be harder to achieve than might be imagined since all too few scientists are completely at home in this type of activity. Judgements must often be made on incomplete information but, in the pursuit of biological control for profit, such progressive development is essential. While the progress in Western Europe is impressive much remains to be done in Southern Europe, the USA and, above all, in the large areas of protected cropping in Japan. Members representing some of these areas should, therefore, consider accepting some guidance from the Working Party, though such stimulus may all too easily be interpreted as interference. Such is the price of international co-operation! However, OILB is largely funded by its national institutional members and the administrative authorities of these bodies would do well to consider very seriously the advantage of assisting this essential development to the benefit both of their growers and consumers.

All must recognize that biological control has largely progressed to its present position through the enthusiasm of a relatively small group of research workers. It is important for this group to avoid duplicating their endeavours. Since most progress over the next decade will largely be in the field of resistant strains of natural enemies and the development of new pathogens, the dangers of overlap increase. The OILB Working Group will, hopefully, play an increasingly important role in assisting these endeavours.

The Group publishes a small cyclostyled newsletter *Sting* from Naaldwijk, which is circulated to several hundred interested workers worldwide. This source of information gives 'advanced notice' of research development and so assists in the general awareness of research staff so that any duplication is, at least, informed.

REFERENCES

Bravenboer, L. ed. (1971) *Proceedings of Conference on Integrated Control in Glasshouses, Naaldwijk* 1970 IOBC/WPRS. 78 pp.

Report of Working Group on Integrated Control in Glasshouses, Littlehampton, UK 1973. *International Organization for Biological Control of Noxious Animals and Plants/West Palaearctic Regional Section Bulletin* **1973** (4): 73 pp.

Report of Working Group on Integrated Control in Glasshouses, Antibes, France (1977). *International Organization for Biological Control of Noxious Animals and Plants/West Palaearctic Regional Section Bulletin* **4** (3): 184 pp.

Report of Working Group on Integrated Control in Glasshouses, Vantaa, Finland (1980). *International Organization for Biological Control of Noxious Animals and Plants/West Palaearctic Regional Section Bulletin* **3** (3): 257 pp.

Report of Working Group on Integrated Control in Glasshouses, Darmstadt, West Germany (1983). *International Organization for Biological Control of Noxious Animals and Plants/West Palaearctic Regional Section Bulletin* **6** (3): 229 pp.

13 THE FUTURE
N.W. Hussey

It is imperative that integrated pest and disease programmes are adapted to changes in cultural techniques and economic circumstances. Since the first programmes were worked out and exploited in the early 1970s, new growing systems have had far reaching effects.

Both the peat-bolster and nutrient-film technique involved covering the soil surface with plastic sheets and so soil sterilization became unnecessary. Or did it? Experience has shown that those pests, or even natural enemies, that spend part of their life cycle in the soil are now able to survive with both positive and negative effects. Noctuid moths, such as *Laconobia*, *Spodoptera* and *Mamestra* spp., are no longer eliminated by autumn hygiene while, as stressed earlier, thrips now abound on the new artificial surface.

One very serious effect of the new 'no-sterilization' regimes concerns agromyzid flies of the genus *Liriomyza*. *L. bryoniae* has become an increasingly serious problem on tomatoes over a large area of Western Europe, whilst the exotic species, *L. trifolii*, is thriving over most of the Continent, despite the fact that some countries eliminate outbreaks by classifying the species as a scheduled pest.

Pupal parasites of this complex of soil pupating insects can also exploit the lack of sterilization so that, following a single introduction, they are able to survive from one season to the next. Braconid parasites of leaf-miners and the aphid-predator, *Aphidoletes*, are able to provide control in successive seasons as they re-appear each new growing season as soon as their hosts appear on the crop.

Energy-saving measures may change the environment and will demand modifications to crop protection programmes. Where thermal screens are used to reduce heat losses, humidities tend to rise, especially in small glasshouses, and while this increases the efficiency of entomophilic fungi, such as *Verticillium lecanii*, the frequency of fungicide application for control of phytopathogenic diseases must also increase. On the other hand, the amount of air exchange is reduced by the screens so that pesticide vapours persist longer, with positive advantage to their primary pest target yet greater deleterious effects on natural enemies.

'Edge' effects are much reduced by the virtual elimination of draughts, so that both pests and their enemies are able to exploit the environment more fully. These trends are partially confounded by lower night temperatures, though where radiant heaters are used the plant surface temperature may actually rise and so increase insect activity.

The other cultural technique liable to rapid change is the irrigation system. Hose-watering has largely been abandoned in favour of 'trickle' systems and various types of overhead watering, which either cover only the soil around the plant stems or, sometimes, the entire plant. Naturally, the effects on the greenhouse environment are dramatic and must be taken into account in pest and disease management programmes.

It is important to recognize that biological control was adopted as a relevant component of pest control technology under glass as a method of preventing, or at least delaying, problems of resistance. It is salutary to reflect that, now that this spectre has largely disappeared as a problem, growers are all too ready to forget the painful experience and revert to the currently effective synthetic pyrethroids, despite the fact that some strains of whitefly are already totally resistant to these compounds. This has been caused by cross-resistance phenomena rather than through the currently limited exposure. It is almost certain, therefore, that the commercial enthusiasm for biological control will be cyclical, with consequent economic constraints on rearing companies.

These companies should, therefore, develop larger markets outside the confines of the greenhouse industry. As greenhouses are largely concentrated in northern latitudes, where outdoor pest problems are few, it is essential to develop biological control technologies for protected and outdoor vegetables in the Mediterranean. The OILB Working Party has already recognized the necessity for this development and is to hold a meeting in Sicily in 1984 in an attempt to stimulate the necessary development work. At least one commercial company is already working in Spain, Greece and Crete, and another in Tenerife. Bearing in mind the importance of the continued viability of the commercial pioneers in biological control, a very strong case could be made for governmental support or bank facilities to increase both the scale and speed of developmental work to ensure a sound technological base for extended commercial exploitation.

My friends, Dr Nakazawa of the Hiroshima Experimental Station and Dr Kajita of Kyushu University, have commented upon the apparent delay in the implementation of biological control within the vast area (32 000 ha) of plastic-protected structures in Japan. While much valuable research has been undertaken and reported, they have yet to develop a practical integrated programme. They have a significant pest, *Thrips palmi*, for which there is, as yet, no effective control, though Thripstick® could provide a solution. Another difficulty concerns the potential dangers of using microbial pesticides in greenhouses in areas where sericulture is practised. Attention has, therefore, remained concentrated on pesticides and, in particular, on buprofezine (Applaud®) which apparently gives control for up to 7 weeks. However, whatever temporary success chemicals have achieved, it seems almost inevitable that resistance in one pest or another will lead that country too towards the biological solution.

However, this rather negative analysis could become less relevant if one, or both, of two other aspects of the problem assume importance.

The first concerns the pesticide manufacturing companies. The future profitability of pesticide manufacture is no longer assured since the massive development costs – in excess of £15 million per product – are reducing the range of products coming on to the market. Some firms are, therefore, considering the sale of complete pest, disease and weed control packages for specific cropping systems. Such packages would, where appropriate, include biological control. To advance these packages, such firms could be prepared to finance the development costs for certain biological products where traditional chemical methods had limitations. An example is cotton where the use of broad-spectrum products for caterpillar control leads to 'rebound problems' with whiteflies, thrips and jassids. So long as the total market exceeds £20 million then such a development should be commercially feasible. Since the advanced stage of integrated programmes for vegetable production would require a much smaller investment, then it is possible that, for firms who already publicize details of biological control techniques in advisory brochures advertising pesticides for greenhouse use, the sale of complete integrated packages would be but a limited further step.

Obviously, it will take time for the boards of accountants and chemists, who largely control the policies of pesticide companies, to accept the rationale of developments directed towards the sale of such packages but the trend could be hastened by pressures coming from their customers – the growers.

There is much public concern about pesticides, mostly fanned by emotion rather than fact, but it is creating an environment in which the mention of non-chemical methods is greeted enthusiastically. This changing attitude by the public is matched by an increasing awareness by growers that intense pesticide programmes significantly reduce the yield and quality of crops. Any reduction in quality is now important commercially as tendencies to over-production in developed countries have created marked premiums for good quality products. Indeed, in Guernsey, 'dumping' of low-grade tomatoes has become an annual event as only top-grade fruit can be sold in summer. Nowhere is this clearer than in ornamental crops where the highest prices are commanded by crops treated with the least pesticide.

So, the challenge is there. What should the International Organization for Biological Control (IOBC) do to build upon the significant advances it has already pioneered?

The emphasis of the Group must move towards the demonstration of complete pest and disease control programmes designed for specific crops in different regions of Europe. The activity of the Working Group concerned with orchards linked with the OILB Commission on the use of 'Trade-marks' to popularize produce which has received minimal pesticide contamination shows the way ahead. One UK company already obtains a market premium by marketing under a 'natural' label. The commercial use of natural enemies in greenhouses has been technically successful but more growers must become aware of its economic advantages and more serious attempts made to advertise the availability of 'naturally-produced' vegetables. The latter is currently a 'gimmick' which could, with OILB support, become an accepted criteria in the market place now that the public is concerned about contaminated food.

From the technical standpoint, there are two areas of potential development. There has been an increased recognition of the value of strains of natural enemies tolerant to different pesticides. Several have already been found naturally as a result of selection pressure in greenhouses and orchards whilst a few have been developed from programmed laboratory selection. Rearing companies should take advantage of these while research units should devise effective selection techniques designed for mass-production units.

Further, in view of their inherent selectivity, more effort should be put into the detection, isolation, production and safety clearance of additional pathogens. Bearing in mind the expensive research and development involved, and it is now recognized that the development needs to be more extensive and so take rather longer than is customary with pesticides, the costs require to be met by firms whose cash flow comes from quite different products. Perhaps, therefore, we should be seeking to 'sell' natural pest control to groups of industrialists with whom we have had no serious contact in the hope that some diversification could open up the possibility of more effectively funded development.

Whatever the problems, it seems clear that biological control has become commercially acceptable and is here to stay, though the methods used may be subject to continuous change.

14 COMMERCIAL PRODUCERS OF BENEFICIAL INSECTS

Biocontrol Ltd, Po Box 515, Warwick, Queensland 4370, Australia (*P.p.*)

Applied BioNomics, Research Station, 8801 East Saanickton Road, Sidney BC, Canada V8L 1H3 (*P.p., E.f.*)
Better Yields Insects, 13310 Riverside Drive East, Tecumseh, Ontario, Canada N8N 1B2 (*P.p., E.f.*)

Kemira Oy, Box 14, 02270 Espoo 27, Finland (*P.p., A.a.*)

CTIFL/SAP, Centre de Balandran, 30127 Bellegarde, France (*E.f.*)

Biological Control Insectaries, Kibbutz Sde Eliyahu, 10810 Israel (*P.p.*)

Koppert BV, Veilingweg 8a, 2651 BE Berkel-Rodenrijs, Netherlands (*P.p., E.f., A.m., L.* p./*O*)

LOG, Okern torg VI, Oslo 5, Norway (P.p.)

Anticimex AB, c/o Tradgardshallen, S-25229 Helsingborg, Sweden (*P.p., E.f.*)

Bunting & Sons, The Nurseries, Great Horkesley, Colchester, Essex, England (*P.p., E.f., A. mat., L.*p./*D.*)
English Woodlands, The Old Barn, Rokelane, Godalming, Surrey, England GU8 5 NT (*P.p., E.f.*)
Humber Growers, Common Lane, Welton, Brough, North Humberside, England (*P.p., E.f.*)
Natural Pest Control, Watermead, Yapton Road, Barnham, Bognor Regis, Sussex, England (*P.p., E.f.,* L.p./*D.*)

Biotactics Ltd, 22412 Pico St, Colton CA 92324, USA (*P.p.*)
Whitefly Control Co., Box 986, Milpitas Ca 95035, USA (*E.f.*)
California Green Lacewings, PO Box 2495, Merced CA 95340, USA (*C.c.*)
Entomological Engineering, J.C. Wagoner Co., Route 2, Box 2410d, Davis CA 95616 USA (*E.f.*)
Organic Pest Control Naturally, PO Box 55267, Seattle WA 98155, USA (*P.p., E.f.*)

Species: *C.c.–Chrysopa carnea*; *P.p.–Phytoseiulus persimilis*; *E.f.–Encarsia formosa*; *A.a.–Aphidoletes aphidomyza*; *A.mat.–Aphidius matricariae*; *A.m.–Amblyseiulus mckenzei*; L.p./*D.–Liriomyza* parasites (*Dacnusa sibirica*); L.p./*O.–Liriomyza* parasites (*Opius pallipes*)

INDEX

Numbers in *italic* refer to black and white illustrations. Those in **bold** refer to colour plates.